GIS Data Conversion: Strategies, Techniques, and Management

Pat Hohl, Editor

Africa • Australia • Canada • Denmark • Japan • Mexico • New Zealand
Philippines • Puerto Rico • Singapore • United Kingdom • United States

NOTICE TO THE READER

Publisher does not warrant or guarantee any of the products described herein or perform any independent analysis in connection with any of the product information contained herein. Publisher does not assume, and expressly disclaims, any obligation to obtain and include information other than that provided to it by the manufacturer.

The reader is expressly warned to consider and adopt all safety precautions that might be indicated by the activities herein and to avoid all potential hazards. By following the instructions contained herein, the reader willingly assumes all risks in connection with such instructions.

The publisher makes no representation or warranties of any kind, including but not limited to, the warranties of fitness for particular purpose or merchantability, nor are any such representations implied with respect to the material set forth herein, and the publisher takes no responsibility with respect to such material. The publisher shall not be liable for any special, consequential, or exemplary damages resulting, in whole or part, from the readers' use of, or reliance upon, this material.

OnWord Press Staff

Publisher: Alar Elken
Executive Editor: Sandy Clark
Acquisitions Editor: James Gish
Managing Editor: Carol Leyba
Development Editor: Daril Bentley
Editorial Assistant: Fionnuala McAvey
Executive Marketing Manager: Maura Theriault
Executive Production Manager: Mary Ellen Black
Production and Art & Design Coordinator: Leyba Associates
Manufacturing Director: Andrew Crouth
Technology Project Manager: Tom Smith
Cover Design by Lynne Egensteiner and Cammi Noah

First edition, 1998
SAN 694-0269

10 9 8 7 6 5 4 3 2

Printed in the United States of America

Library of Congress Cataloging-in-Publication Data
GIS data conversion: strategies, techniques, and management / Pat Hohl, ed.
p. cm.
Includes index.
ISBN 1-56690-175-8
1. Geographic information systems. I. Hohl, Pat, 1961– .
G70.212.G572 1997
910'.285—dc21 97-45483
CIP

For more information, contact
OnWord Press An imprint of Thomson Learning
Box 15-015 Albany, New York USA 12212-15015

About the Editor

Pat Hohl is a senior electric engineer with the City of Riverside (California) Public Utilities Department. Pat earned a B.S.E.E. from Drexel University in 1984. He has served as the city's CADME/GIS project manager responsible for data conversion, software development, and implementation. Pat also co-authored the ArcView GIS Exercise Book (OnWord Press 1996, 1997). He is an instructor for the University of California-Riverside UNEX GIS certificate program, and president of the Inland Empire URISA Section. On a lighter note, Pat and his dog Trooper are Regional Division II Canine Frisbee champions.

Information about other contributors can be found in Appendix B, "Contributors."

Acknowledgments

I would like to thank all of the people who helped to make this book possible. Thanks to my lovely wife Dena for her love, support, and encouragement, and little Curtis for his *help* at the computer. I extend my sincere gratitude to each of the contributors. I have never stuck my neck out so far, relying on a large group of individuals all over the world, and been so pleased with the outcome. Thanks, gang! I hope we can work together again. Finally, I'd like to acknowledge the team at OnWord Press for their cooperation and assistance throughout the entire project.

Pat Hohl

Image Credits

The editor and publisher are grateful for image contributions throughout the book. Images contributed by chapter authors are not credited. When credit is known, images are attributed using the following abbreviations, which appear in individual image captions.

- Section pages and opening pages of chapters. MSE Corporation.
- Chapter 5. William E. Fagan (WEF).
- Chapter 6. Mike Rottler-Gurley, Christopher Thomas, and the Environmental Systems Research Institute, Inc. (ESRI).
- Chapter 7. MSE Corporation (MSE).
- Chapter 9. Rocky Mountain Mapping Center of the U.S. Geological Survey and EROS Data Center Distributed Active Archive Center (RMMC).

- Chapters 10 and 11. Ray McCollum (RM), U.S. Army Corps of Engineers (USACE); Dario Franzi (DF), USACE; and Photo Science, Inc. (PS).
- Chapter 12. Trimble Navigation, Ltd. (all images).

Typographical Conventions

The following shorthand conventions are used to identify distance measurements: 30" (30 inches); 25' (25 feet); 6.2 mi (6.2 miles); and 1/4" tape. Metric distances are abbreviated as follows: 2.5m, 130km, .3cm in length. Angle degrees are spelled out unless five or more references occur in a chapter. All measurements are approximate unless otherwise noted by the editor.

File formats, Web addresses, and other information that a user must type are *italicized*.

ARC/INFO Commands

ARC/INFO command prompts described in Chapter 15, "Spatial Data Transfer Standards," are CAPITALIZED in full. Statements the user must type are *italicized*. Reports generated by ARC/INFO appear in `courier` type.

Web Updates

Readers curious about updates or corrections to this book may access the Web site below.

http://www.onwordpress.com/

- Chapter 12. Trimble Navigation, Ltd. (all images).

Typographical Conventions

The following shorthand conventions are used to identify distance measurements: 30" (30 inches); 25' (25 feet); 6.2 mi (6.2 miles); and 1/4" tape. Metric distances are abbreviated as follows: 2.5m, 130km, .3cm in length. Angle degrees are spelled out unless five or more references occur in a chapter. All measurements are approximate unless otherwise noted by the editor.

File formats, Web addresses, and other information that a user must type are *italicized.*

ARC/INFO Commands

ARC/INFO command prompts described in Chapter 15, "Spatial Data Transfer Standards," are CAPITALIZED in full. Statements the user must type are *italicized.* Reports generated by ARC/INFO appear in `courier` type.

Web Updates

Readers curious about updates or corrections to this book may access the Web site below.

http://onwordpress.com/updates.cfm

OnWord Press...

OnWord Press is dedicated to the fine art of professional documentation.

In addition to the editors and contributors who developed the material for this book, other members of the OnWord Press team contributed their skills to make the book a reality. Thanks to the following people and other members of the OnWord Press team who contributed to the production and distribution of this book.

Dan Raker, President
Dale Bennie, Vice President, Publishing
Rena Rully, Publisher
Carol Leyba, Associate Publisher, Books and Software
Barbara Kohl, Associate Editor, Books and Software
Cynthia Welch, Production Manager, Books and Software
Jean Cooksey, Development Editor
Lynne Egensteiner, Cover designer, Illustrator

Contents

SECTION 2: Project Management 27

Chapter 3: Conversion Resources and Structures 29

Chapter 4: Project Planning and Management 63

Chapter 5: Documenting a Data Conversion Effort 99

SECTION 3: Understanding the Target System 113

Chapter 6: Understanding the Target System 115

Chapter 8: Data Models, Collection Considerations, and Cartographic Issues 161

Chapter 9: External Data Sources and Formats 179

SECTION 5: Data Conversion/Input Methodologies 205

Chapter 10: Airborne Sensing Systems and Techniques 207

SECTION 6: Quality Control/Quality Assurance 355

Chapter 16: Quality Control/Quality Assurance 357

Appendix A: Glossary 383

Appendix B: Contributors 393

Index 397

Section 1: Introduction

This section introduces the issues involved in data conversion for geographical information systems (GISs) and explains how data are incorporated into conversion projects.

- **Chapter 1** provides an overview of the book and explains the topics covered in each section.
- **Chapter 2** defines data in the context of conversion, differentiates among basic data types, discusses how data are processed, and provides a brief introduction to the relational data model, a foundation of GIS database organization.

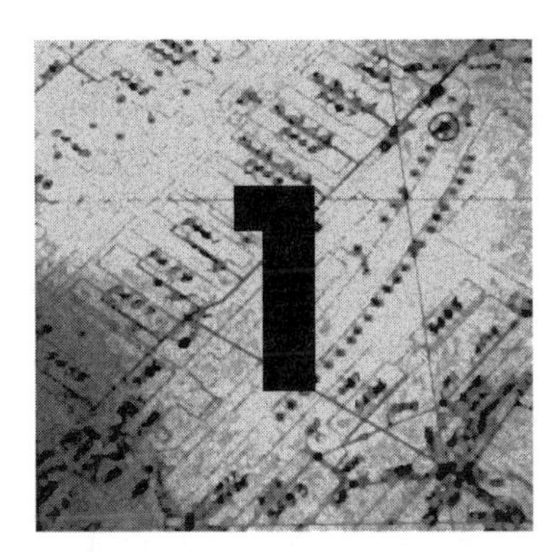

Introduction and Overview

Pat Hohl

The foundation of every geographic information and automated mapping system is the database. The process of creating a database from source information is commonly called *data conversion.* Almost without exception, source data must undergo some form of conversion in order to support your intended applications. Therefore, data conversion is essential before you can realize anticipated benefits from your system. Conversion may be from existing hardcopy or digital records or from new data collection, such as aerial photography or a field survey.

As a rule of thumb, data conversion typically accounts for 50 percent of the cost and effort of a GIS project. The conversion phase is usually a onetime event in the life cycle of a GIS project or program. If your conversion is fully one half of the entire program and it falls short, will you get a second chance?

Ensuring Successful Conversion

If you only have one opportunity for conversion, it should be the best effort you can put forth. Therefore, it is worthy of careful study, planning, and execution. This book will show you how to navigate the process. It takes much more that a digitizer to create a useful GIS database. A thorough understanding of the process should help you achieve a higher quality end product. High quality data should have a longer service life, support better analysis, and have greater potential for resale or barter.

The source data may be in many different forms, such as hardcopy maps, microfiche, photographs, notebooks, or digital records. In addition, the project may entail converting analog data to digital form. The process may range from a simple import supported by your GIS software, or it may be very complex, such as processing custom aerial photography to locate features of interest. This book explores the pertinent issues affecting GIS data conversion, including management of the process, various approaches, resource allocation, data sources, hardware/software requirements, and specific conversion methodologies. GIS hardware and software change quickly, but the fundamental concepts of data conversion remain relatively constant. A handy reference guide, this book will help educate you regarding relevant issues, technologies, and trends.

Book Audience

This book provides an overview of the GIS data conversion process for individuals involved in such projects. It also addresses many of the technical issues independent of specific GIS platforms and database designs. Managers, technicians, and decision makers alike will find the presentation of related issues useful in planning and executing a conversion effort. Properly converted data will provide a solid foundation for GIS analysis enabling you to realize the anticipated benefits of the system.

Historically, some projects have failed to meet expectations because of their failure to deal with conversion issues. What works in one set of circumstances may not work in a different set of circumstances. There are no easy cookbook solutions to data conversion. It is labor-intensive and not very glamorous. However, conversion can also be quite interesting as a result of the numerous technologies involved.

Who, What, Why, Where, and When?

Hardware and software change, but data persist. Many experts agree the core value of a GIS is the data. Users must be able to retrieve solid answers about the origin and limitations of the data to use them effectively and avoid using them ineffectively.

Your organization will undoubtedly implement GIS with specific analysis output in mind. To realize the benefits of a GIS, several components are required, including hardware, software, appropriate algorithms, and suitable data. As noted previously, the data form the groundwork upon which the entire system rests.

What would you do if asked by your boss, board of directors, or city council to validate the results of your GIS analysis? If you had to defend your GIS analysis in a court of law, an opposing attorney would attack the analysis algorithms—and, more importantly—the data on which they are based. This scenario leads to several questions, any of which you might be required to answer (whether in a court of law or business situation).

- Who created the data? What training, skills, and experience do they have?
- What exactly do the data depict?
- Where did the data originate? What was their source? Did a previous source exist?
- Why were the data created in this particular way?
- If the data were created for 1" = 1,000' mapping, are you trying to measure the width of a sidewalk with them?
- When were they created? Have they been updated?

The nuts and bolts of data origins and processing must be thoroughly evaluated in a GIS data conversion effort. The same issues apply to ongoing maintenance as well; edits to the master database can be viewed as minor data conversion efforts.

Because the data conversion process can often be lengthy, you can help yourself and the project by demonstrating its benefits early on to help maintain support. As such, you should determine in advance how to illustrate its benefits throughout the process. Use early data to achieve something useful. Convert a set of data quickly, so that you can demonstrate the advantages of your efforts. One city undertaking a very thorough data conversion process converted streets quickly because they were fairly straightforward. The streets were used to promptly replace an aging city-wide map book while years of additional facilities conversion took place. This simple application demonstrated early benefits and bolstered support for the project during the labor-intensive conversion.

Different Perspectives

The fundamental principles of project management, data source assessment, selection of optimal techniques, and quality control/quality assurance affect all GIS data conversion projects. Data are converted to support various applications, and these applications drive GIS systems. The application drivers of GIS projects around the world are as diverse as the people using them. As a result, a broad spectrum of GIS data conversion and accuracy requirements exist in the industry.

Consider the different perspectives of a small GIS shop compared to a large government agency. Large agencies (with correspondingly large budgets) typically have numerous workstations, large geographic extents, large staffs, and ± 40' (or less) accuracy requirements. Business marketing applications might require an approximate location, such as ± one city block, while delivery routing applications would require data specific enough for a truck driver to find a customer's address.

Engineering or utility users typically create 1" = 40' map products with ± 2' accuracy or better. Because public safety and utility reliability depend on precise data, updates may be made to the database the same day a change is made in the field. Approaches to GIS data conversion range from work done in house over a long period of time to fast-track conversions accomplished by highly skilled vendors. Obviously, there are many different perspectives on data and conversion for GIS applications.

Book Organization

This book is a compilation of papers by experts in various fields, each with different experience. These experts have different viewpoints, and some overlap naturally exists from topic to topic. A broad range of information is presented in order to address the many forms GIS data conversion can take, and to give you the information you need to assess your own GIS data conversion needs. The book is divided into the following sections.

- Introduction and overview
- Project management
- Understanding the target system
- GIS data sources
- Data conversion/input methodologies
- Quality control/quality assurance

Introduction and Overview

As noted before, data are the very foundation of a GIS. Data have various characteristics, such as precision and accuracy, and are of distinct types (e.g., raster and vector). A thorough understanding of the nature of the data to be converted is critical to effective conversion for GIS use.

Project Management

Project management of GIS data conversion is similar to good project management in general. For example, clear objectives, good communication, and identifiable milestones are part of any successful management effort. However, several unique elements may be unfamiliar to even seasoned managers. Many strategic issues come into play for the majority of conversion projects. An appropriate pilot phase, minimal scope creep, and effective risk mitigation can make the difference in your conversion. Consider the logistics of handling every source document, and converting and tracking the documents—all without adversely impacting the organization's work routine—and you gain a sense of the complexity.

Management of data conversion, like so many things in life, quickly becomes an exercise in balancing trade-offs. Normally, you can only have two of the three coveted conversion characteristics—good, fast, and cheap—and must sacrifice the third. You can have data fast and cheap, but they're not going to be the best possible quality. You can obtain good data quickly, but they will be expensive. Or, you can have good data cheaply, but you're not going to get them very fast. This section also focuses on how to effectively document conversion requirements so that they are clearly understood.

Data conversion trade-offs.

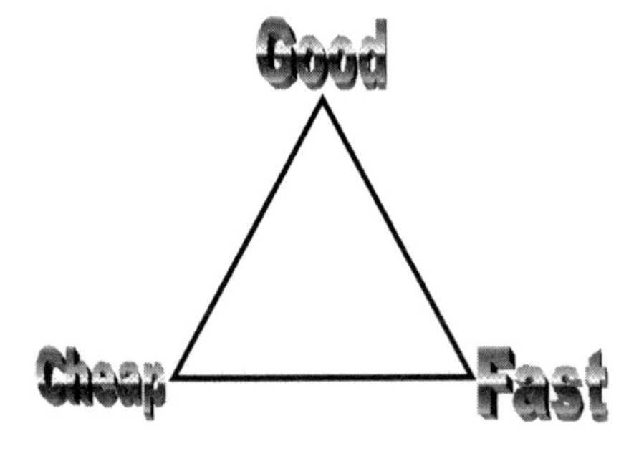

External expertise is frequently required to meet the schedule, budget, and/or requirements of your data conversion effort. For example, it is unlikely you would secure the trained staff and equipment to perform stereocompilation of aerial photographs within your organization; most people hire an expert vendor to accomplish this. There are other areas where you may need assistance. Consultants and data conversion vendors are critical to successful conversion in many cases. This section will show you how to evaluate their qualifications.

Understanding the Target System

A clear understanding of the target system (where the data will be used) is essential to meet your conversion objectives. Data organization and storage interact with your system's hardware, operating system, and network. Careful consideration must be given to the benefits and security risks of the target system's network environment and use of the ubiquitous Internet.

GIS Data Sources

If data are the foundation of your GIS application, conversion sources constitute the foundation of the data. Data sources are critical to the conversion process, and the accuracy and quality of the resulting data are limited by their sources. Existing and new data sources must be carefully evaluated and prepared for the process of conversion. These sources can take many forms—from wall maps to 3 x 5" index cards. In addition, aerial photography, satellite imagery, or field inventory may be used to supplement existing data. Attributes and graphics often derive from different sources and must be synthesized.

Knowledge of data sources is key to using data effectively. Because this knowledge may exist only among (or within) individuals in your organization, personnel issues are also very important to the success of your project. This section helps you to evaluate the strengths and weaknesses of available sources and select the optimal mix.

Data Conversion/Input Methodologies

Numerous techniques and methodologies are available to actually convert source data into the target digital format. Each technique has specific advantages and limitations. How you select optimal techniques depends on source data, resources, and objectives. Will you use digitizing, photogrammetry, satellite imagery, or perhaps COGO? If you plan to conduct a field inventory, GPS may fit in the mix. If your system will utilize raster data, you will want to explore scanning and related technologies. If you intend to use feature attributes, you will need to consider keyboard entry of the values. This section also investigates spatial data transfer standards, which may become important if you plan to exchange data with others.

Quality Control/Quality Assurance

Regardless of the conversion techniques you use, you must ensure an appropriate level of quality in your converted data. Terms such as data quality and accuracy are very context sensitive. In one case, accuracy of a line may refer to its graphical properties, such as how many vertices it uses to represent the curvature of a line. In another case, accuracy of a line might refer to how closely coordinate end points represent their true locations on the Earth's surface. In a third case, attribute inaccuracies might cause a street to be flagged as a river.

What is data "quality" anyway? What is the difference between quality control and quality assurance? This section is focused on these questions, as well as procedures for checking data and establishing acceptance standards. You must determine the level of quality you really need, how to achieve it, and how to verify that the conversion process is meeting the standard.

Summary

The adage "plan your work and work your plan" certainly rings true for GIS data conversion. If you accurately determine your true needs, select optimal sources and conversion techniques, monitor data quality, and manage the process effectively, the result will be a database in which you can have confidence. These data will be the best data for your applications and should form a solid platform for your system.

Data, the Foundation of GIS

Andrew Coates

Geographic information systems (GISs) have enabled information analyses that were not possible even 20 years ago. However, in order to perform these analyses, you must have something to study. In GIS, this takes the form of data. This chapter examines the building blocks of GIS. The discussion begins with an explanation of what data are and the types of data used in a GIS. The critical concepts of precision and accuracy are introduced. Data sources and the types of information that can be pulled from data are explored. Finally, the relational data model is briefly discussed. A working knowledge of these concepts is essential to effective data conversion.

Data Definitions

Information versus Data

Although the terms information and data are often used synonymously, they are not identical. Data are what you collect through observation, measurement, and inference. Only after you analyze and organize masses of data can you produce information. Therefore, think of information as data in a useful form. The main role of a GIS is to convert data into information.

For example, a retail chain may collect data about where its customers live and what they purchase when visiting one of the chain's stores. This may result in a list of postal codes and purchase amounts, an example of which appears below. Of course, these data are generally not useful until they have been analyzed—or, in the case of GIS, plotted on a map. Transforming these data (see the map following the list) indicates that the majority of purchases are made by people from a particular area. The data have been converted into information that can be used to more precisely target the next marketing campaign.

Postal codes and purchase amounts.

```
53960 125.88 53960 124.47 53960  38.49 53960 185.56
53960 189.51 53960   8.14 53960 187.80 53960  48.74
53960  29.10 53911  15.54 53911  23.61 53583  80.96
53583  58.18 53583  41.56 53583 135.04 53583  48.52
53529  75.45 53529  90.65 53529 160.90 53529  42.26
53529  50.92 53529  94.59 53529  86.57 53529  86.81
53555  71.56 53555  74.30 53555  75.05 53555 142.31
53555  75.86 53925 122.63 53925 100.22 53925 158.98
53925  59.40 53925  60.68 53925 162.42 53590 131.04
53590 184.09 53590  37.69 53590 190.40 53590 160.56
53590  46.17 53590  48.85 53590 111.50 53532  15.06
53532  37.58 53532 198.90 53532  34.80 53532 157.84
53532  52.41 53597  39.53 53597  67.73 53597  22.67
53597 133.80 53597 171.01 53597 100.08 53559  43.58
53559  22.77 53559 132.77 53598  43.92 53598 121.56
53598  16.69 53598 101.19 53598  70.63 53598  93.03
53598 131.85 53598  80.74 53560 146.46 53560 114.75
```

The data have been transformed into information using GIS.

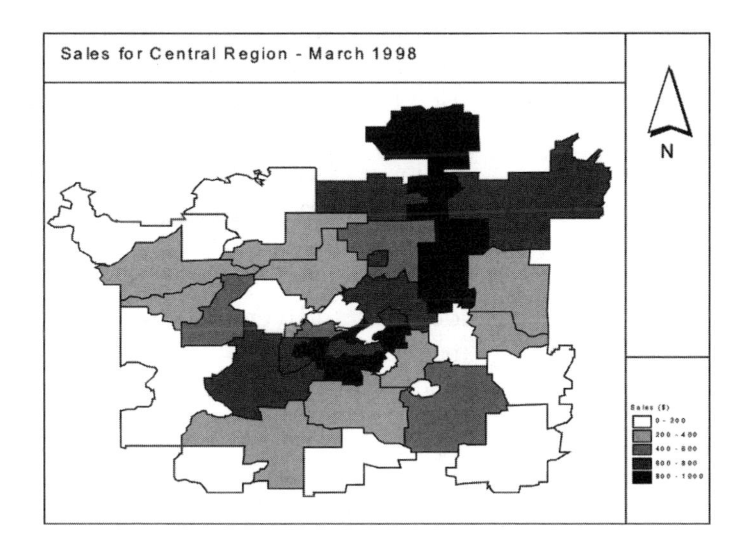

Precision versus Accuracy

When referring to data, two other terms—precision and accuracy—are often used interchangeably, but they also mean different things in this context. The precision of data relates to how they are stored, often reflecting the number of significant digits to which a number is quoted. Precision is quite simple to quantify.

On the other hand, the accuracy of data refer to how well they were measured, recorded, stored, converted, manipulated, transformed, and presented. Quantifying this characteristic is far more difficult. Data accuracy is an extremely important concept in any data set. In order to make sensible decisions based on your data, you must know whether they provide an exact (highly accurate) or approximate (less accurate) picture of reality.

✓ **TIP:** *While you should always store data with as much precision as possible, it is also important to maintain careful notes in the metadata (see the "Metadata" section below) about the level of data accuracy.*

Less accurate data are not necessarily less useful. In the example of the retail chain, the centroid of each suburb is used to represent the location of all customers in that suburb. While this is only an approximate representation of reality, it is sufficient for the purpose of a proposed marketing campaign. However, if door-to-door sales staff were to visit each customer, the data would not be sufficiently accurate. The next figure shows an example of regions with locations and very precise coordinates for associated centroids.

Low accuracy information displayed with high precision.

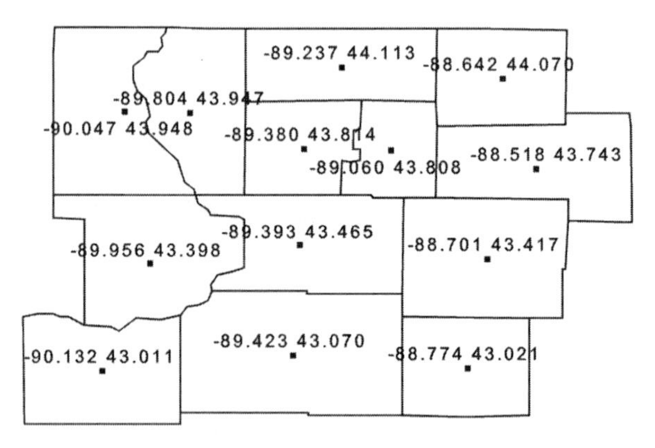

For a more detailed discussion of precision and accuracy, see Chapter 8, "Data Models, Collection Considerations, and Cartographic Issues."

Data Types

Data can also be classified on other grounds, such as collection method. For example, physically measured data (e.g., the radar based reflectivity of a corn field) can be categorized separately from inferred data (e.g., soil salinity in the same field). You will probably use several different classification schema in your work with geographic data.

For many modeling scenarios, data from different classifications are required. Modeling a river catchment effectively requires spatial data on catchment boundaries, slope, land use on the catchment, soil type, the drainage network location and slope, and the location of various measuring stations. It also requires temporal data, such as hydrographs of recorded storm events to calibrate the data model, and rainfall patterns at the rain gauge stations for calibration and modeling events. Finally, it requires attribute data such as infiltration parameters for each soil type, the percentage impervious area for each land use, and characteristics of each element of the drainage network (roughness, slope, and size).

Spatial Data

Spatial data record relationships among and about geographically distinguishable features. Examples include the location of a rain gauge, the route a delivery truck takes, the extent of damage from a brush fire, and the center and size of each pixel on a satellite image. Spatial data are generally used in one of two formats. Vector format records spatial data as a series of points, lines, and polygons. Raster format records spatial data as the location and size of each element in a (usually square) grid. See Chapter 8, "Data Models, Collection Considerations, and Cartographic Issues," for more detail on raster versus vector data.

Points

Points (or nodes) are the fundamental building block of spatial data. They refer to a specific place in a generally two-dimensional space. Points are used to record the location of objects (an oil well or telephone box) and form the basis of vectors, polygons, and rasters. In a two-dimensional coordinate system, points require two values to be located. If the coordinate system is Cartesian (e.g., Universal Transverse Mercator projection), the values are distances from an origin in two orthogonal directions. If the coordinate system is spherical (e.g., latitude and longitude), the values are angles offset from a datum in the center of a sphere of known radius, and the points are locations on the surface of the sphere.

Vectors

Vectors are lines between points and are generally stored as an ordered series of two or more nodes. GIS also stores information on vector direction and length. Vectors are used to represent essentially linear features, such as pipelines, roads, and rivers.

Polygons

Polygons are closed vectors, where the first and last node in the series are the same point. GIS also stores polygon area and centroids, along with the list of nodes comprising the polygon's perimeter. The centroid is an additional point that uniquely identifies each polygon. Polygons are used to represent features that have extent (e.g., lakes, political boundaries, and soil type).

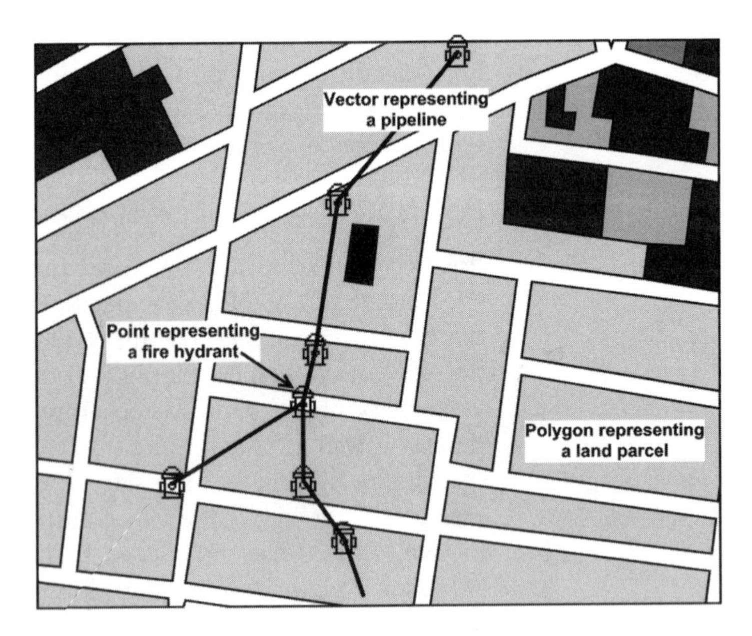

Spatial data in vector format showing points, lines, and polygons.

Rasters

Raster data use a fixed grid and record information about each element of the grid. Points are used to refer to a specific part of each grid cell (often one of the corners). One or more attributes are associated with each element. Raster data, which often come from remote sensing, can include satellite radar images or scanned maps.

Attribute Data

Attribute data generally record information about the objects represented in the spatial data, such as material used to construct a pipe, the type of rain gauge used to measure rainfall intensity, or the percentage impervious area of a particular land use.

Temporal Data

Temporal data record information about how a parameter changes over time. Examples of temporal data include the intensity of rainfall at a gauge, the flow in a stream or pipe, or the evaporation rate in a particular region. For a more detailed discussion of time series data in GIS, see *Time in Geographic Information Systems* by Gail Langran (Taylor & Francis, 1993).

Metadata

One very important data type is metadata, or data about data. Metadata allow the transfer of important comments about the data along with a data set. Information such as the method of data acquisition, the source of attributes recorded with the data, and similar details are all essential for sensible data usage.

Data Sources

You are probably reading this book because you have some data you must convert from one format to another. To make sensible data conversion decisions, you must understand the origin of the data you plan to convert. This section explores the various origins of data and relevant issues affecting each data source.

Measured Data

Physically collected data, such as surveyors determining the location of a pipeline, are referred to as measured data. As long as the measurement process is well understood and thoroughly checked, this source is generally the most reliable. When checking measured data, it is critical to identify all potential sources of error because something as simple as transcribing field notes incorrectly can greatly diminish data usefulness.

Inferred Data

Data calculated from other data are known as inferred data. One example would be the types of crops in a given field, which are inferred from the electromagnetic radiation reflected from the field. When using inferred data it is important to understand the source of the base data underlying the inferences, as well as the model used to perform the calculations. Both factors influence accuracy and reliability.

Imported (and Converted) Data

It is rare that all data used in a decision making process come from the same source or even the same organization. Rather, data are usually imported into or converted from disparate sources. When using such data, you must fully understand the process used to perform the conversion, the original source of the data, and the data manipulation carried out on the data. For this reason, metadata are critical because they transmit vital ancillary information with the actual data set.

Data Processing

There are many stages in the process of creating useful information from data. In *Integrated Environmental Monitoring and Information Usage in Catchment Management* by J.E. Ball, A. Coates, and T.D. Waite (Proceedings of the Asian Industry Technology Conference, 1997) lists the following stages in data collection and management: sensing, recording, transmission, translation, editing, storage, and retrieval. In addition, analysis and presentation are necessary to transform the data into a form useful for decision makers. This generic data life cycle is illustrated below.

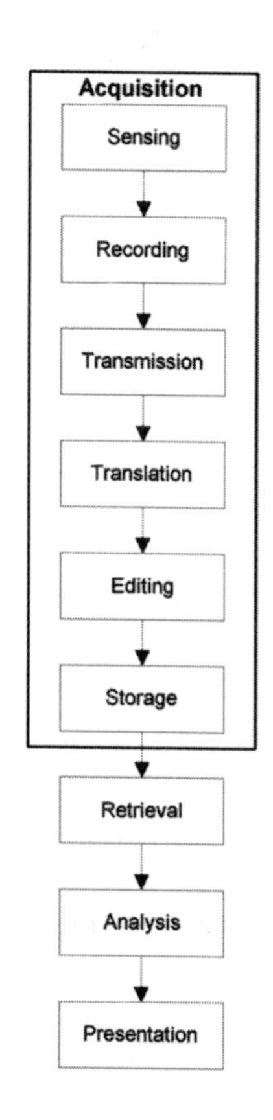

Stages in the data life cycle.

Acquisition

The first six stages of the data life cycle can be conceptualized as a self-contained process referred to as acquisition. These stages are required to convert the data into a form suitable for retrieval, analysis, and presentation.

Retrieval

Even when data are stored in some manner (such as in a GIS), they are of little use until retrieved. Various querying options are available both within the storage packages and as external processes. Data retrieval can be carried out with a number of tools, the most common of which is a database. This will usually allow queries to be constructed and data to be retrieved based on certain criteria. Modern GIS systems excel in this area.

Analysis

Data analysis is an important part of the decision making process. The wide range of analysis tools available to the decision support system developer includes statistical analysis packages, numerical modeling tools, and heuristics based analysis tools such as expert systems and neural nets.

Presentation

Data presentation tools facilitate data analysis and interpretation. The tools selected for use depend on the type of analysis performed and the desired output. Graphical and mapping packages (including GIS) are among the tools available for this stage.

Converting, Exporting, and Importing

Each tool available for use within these stages will have distinct data formatting requirements. Data conversion is typically required to transform the data from an output format indicative of one stage for use in a subsequent stage. Although you may find yourself responsible for every stage in the process, such a situation is the exception rather than the rule. Typically, different departments—or even organizations—are separately responsible for acquisition, analysis, and presentation of a single data set. As such, understanding how to import and export data among diverse formats is critical.

Relational Data Model

GIS frequently employs relational database techniques to store, retrieve, and analyze tabular data. While a complete explanation of the relational data model is beyond the scope of this book, a brief discussion is in order. For more detailed information, consult one of the many excellent textbooks on the subject.

Entities

An *entity* (often represented by a table) is a person, place, event, object, or concept about which you want to maintain data. There is an important distinction between entity *types* (i.e., the structure of the table or columns) and entity *instances* (i.e., the rows in a table). Entity types define a "blueprint" for the entity, specifying attributes (or properties) contained in the entity, but not the values assigned to the properties. For example, the entity *car* may contain the attributes *color*, *transmission*, *VIN* (vehicle identification number), *doors*, *licenseno*, *engine capacity*, and so forth. An instance of the car type might have the attribute values *red*, *automatic*, *RJ458/547-87KL*, *3*, *OUN343*, *1800*, and so forth.

Every entity type has an attribute or set of attributes that uniquely identifies each instance, collectively known as a *candidate key.* Some entities have more than one candidate key. In the previous example, both *VIN* and *licenseno* uniquely identify each car. A *primary key* is the candidate key selected as the identifier for an entity type.

Relationships

A model is a set of related entities, and a relationship is the structure defining how the entities in the model interact. A relationship is an association between the instances of one or more entity types. For example, a student records model stores information about the entities *student* and *subject.* To record each student's results for each subject, the model requires a relationship to link the entities. The relationship itself may have attributes (in this case, the *results* relationship contains the attributes *session* and *result*).

Normalization

What creates a well-defined relationship? It should contain all data with minimum redundancy, such that each row of a table contains information about a single entity instance, and each entity instance is represented by a single row of the table. Redundancy can lead to insertion, update, or deletion anomalies. Although many entity relationship sets can be arranged to minimize redundancy intuitively, the most consistent results arise from the normalization process.

Normalization is a set of well-defined steps to reduce complex relationships to simple, stable data structures. Although there are six logically sequential forms into which data can be organized, normalizing to the *third normal form* is sufficient in most cases. Relationships have the properties listed below.

1. *Entries in columns are in their simplest, most basic form.* There are no multivalued entries at any intersection of a row and column.

2. *Entries in columns derive from the same domain.* All entries in a column describe the same attribute of separate entity instances.
3. *Each row is unique.*
4. *The sequence of columns is insignificant.*
5. *The sequence of rows is insignificant.*

The *first normal form* (1NF) consists of removing repeating groups to ensure there is a single value at the intersection of each row and column. For example, employee records maintained by a human resources department may have entries such as those shown in the first table below. By creating a row for each instance of the data, the data are placed in 1NF (see the second table below).

Beginning data set

EMPLOYEENO	NAME	DEPARTMENT	MANAGER	COURSE	DATECOMPLETED
123-456	Johnson	Payroll	987-154	Excel ArcView	17-Feb-1994 16-Jul-1993
654-987	Gleeson	IT	024-024	C++ FoxPro	01-Apr-1995 17-Jul-1991
587-882	Skinner	IT	024-024	VB	12-Dec-1994

1NF data

EMPLOYEENO	NAME	DEPARTMENT	MANAGER	COURSE	DATECOMPLETED
123-456	Johnson	Payroll	987-154	Excel	17-Feb-1994
123-456	Johnson	Payroll	987-154	ArcView	16-Jul-1993
654-987	Gleeson	IT	024-024	C++	01-Apr-1995
654-987	Gleeson	IT	024-024	FoxPro	17-Jul-1991
587-882	Skinner	IT	024-024	VB	12-Dec-1994

In the *second normal form* (2NF), any partial functional dependencies are removed. A relationship in the first normal form will also be in the second normal form if any of the following apply:

1. The primary key consists of a single attribute.
2. No non-primary key attributes exist in the relation.
3. Every non-key attribute is fully functionally dependent on the full set of primary key attributes.

The relationship in the preceding table is not in 2NF because the primary key (a combination of EMPLOYEENO and COURSE) is not required to define all non-key attributes. The solution is to decompose the relationship into two tables in 2NF. The relationships should be EMPLOYEE (EMPLOYEENO, NAME, DEPARTMENT, MANAGER) and COURSES (EMPLOYEENO, COURSE, DATECOMPLETED).

Examples of employee data converted to 2NF are shown in the following two tables.

2NF employees

EMPLOYEENO	*NAME*	*DEPARTMENT*	*MANAGER*
123-456	Johnson	Payroll	987-154
654-987	Gleeson	IT	024-024
587-882	Skinner	IT	024-024

2NF courses

EMPLOYEENO	*COURSE*	*DATECOMPLETED*
123-456	Excel	17-Feb-1994
123-456	ArcView	16-Jul-1993
654-987	C++	01-Apr-1995
654-987	FoxPro	17-Jul-1991
587-882	VB	12-Dec-1994

In the *third normal form* (3NF), transitive dependencies are removed. Transitive dependencies are relationships between non-key attributes within a table. In the example above, each DEPARTMENT has one and only one MANAGER, and each MANAGER manages one and only one DEPARTMENT, so there is a transitive dependency. The solution is to decompose the relationship even further so that they become EMPLOYEE (EMPLOYEENO, NAME, DEPARTMENT), DEPARTMENT (DEPARTMENT, MANAGER), and COURSES (EMPLOYEENO, COURSE, DATECOMPLETED).

There is occasionally some confusion between partial functional and transitive dependencies. Partial functional dependencies occur when the entire primary key is not required to uniquely identify a part of a row or table. An example of this is the employee details and the courses they have completed, shown in the preceding table "1NF data." An employee's details are the same regardless of the course this row is referring to. Therefore, those details are only partially functionally dependent on the primary key for that table.

Transitive dependencies occur when one column in a row can be predicted by knowing the value of another column in the same row, and vice versa. For example, if each department has only one manager and each manager manages only one department, there is a transitive dependency between the manager and department columns. If you know the manager, you also know the department, and vice versa. To store both pieces of information for every employee is redundant. It is more efficient to create a separate managers table in which the manager for each department is stored only once, and to store only the employee's department in the employee table.

Examples of the employee data converted to 3NF are shown in the three tables that follow.

3NF employees

EMPLOYEENO	NAME	DEPARTMENT
123-456	Johnson	Payroll
654-987	Gleeson	IT
587-882	Skinner	IT

3NF departments

DEPARTMENT	MANAGER
Payroll	987-154
IT	024–024

3NF courses

EMPLOYEENO	COURSE	DATE COMPLETED
123-456	Excel	17-Feb-1994
123-456	ArcView	16-Jul-1993
654-987	C++	01-April–1995
654-987	FoxPro	17-Jul-1991
587-882	VB	12-Dec-1994

Conclusion

To gain the most from your conversion process, it is essential that you understand the fundamental components of your data. This chapter has explored the differences between data and information and the concepts of precision and accuracy, and it has introduced common data sources. Finally, a brief discussion of the relational model provided insight into how to organize data for minimum redundancy and maximum efficiency. Armed with this knowledge, you can understand and interpret data from a variety of sources and, more importantly, convert them to a useful form.

References

Ball, J.E., A. Coates, and T.D. Waite. 1997. *Integrated Environmental Monitoring and Information Usage in Catchment Management.* Hong Kong: Proceedings of the Asian Industry Technology Conference.

Date, Chris J. 1994. *An Introduction to Database Systems (6th ed.).* Menlo Park, California: Addison-Wesley.

Langran, Gail. 1993. *Time in Geographic Information Systems.* London: Taylor & Francis.

Section 2: Project Management

This section focuses on overall management of a data conversion effort and offers guidelines for creating useful documentation.

- **Chapter 3** addresses conversion resources and structures. Topics include personnel, budgeting, hardware and software considerations, data sharing opportunities, incremental versus onetime conversion, and pilots.
- **Chapter 4** discusses project planning and management. It includes suggestions for gaining user and management acceptance; evaluating consultants versus in-house projects; limiting change control; and other soft issues that can influence data conversion projects.

- **Chapter 5** discusses the benefits of documentation and suggests four guiding principles: simplicity, portability, motivation, and detail. The elements of effective documentation are covered at length.

Conversion Resources and Structures

Kelly M. Dilks and Kevin Struck

Staffing Needs

Staffing requirements vary for every data conversion project. Your staff requirements will depend upon the scope of the project, milestones required for delivery, the amount of data, and available funding. Ideally, the experts necessary for your project already work within your organization. However, this is rarely true in today's workplace. Staff may need to be retrained or have skills expanded to support the project. In selecting staff, choose personnel who believe in the project. Believers will do everything in their power to help the project succeed and are often more aggressive seeking the help they need to fill knowledge gaps. When hiring additional staff is not appropriate, you may also need to evaluate the use of consultants and vendors.

Project Manager

First and foremost, your project will require a skilled manager you can depend on to enforce budgets and deadlines.

Project managers often bring years of valuable experience to the task in meeting milestones and budgets. They can assess potential solutions and make decisions regarding appropriate courses of action in order to maintain project integrity. Project managers can reallocate funds and personnel when necessary. They are in a position to ask for additional funds and personnel to meet critical deadlines.

Project Champion

"The project must have a leader. By that I mean it can't have no leaders and it can't have two leaders or a committee of leaders. THERE MUST BE ONE LEADER .... When I say 'leader,' I mean not so much the person with the title, but a person who is going to get the project done. She lives, eats and breaths [sic] the project. She is going to get it done or die in the attempt."—Fergus O'Connell, *How To Run Successful Projects* (Prentice Hall, 1994).

Conversion projects also require champions. Depending on the structure of your project, you may fill both roles. You can act as a champion only if you sit high enough in the organization to provide executive level political and financial support in the form of cash, equipment, and personnel. You can fill both roles only when you have the time and proximity to administer project operations and a sufficiently technical background. If you cannot fill both posts—but you cannot determine which role you fill—a struggle may develop to lead the project. This is a battle the champion is bound to win, and the sooner the manager accepts this, the better.

Project managers and champions must exhibit the characteristics below.

- Excellent reading comprehension skills to keep pace with changes in GIS technologies and industry trends.
- The ability to form innovative solutions from complex technical information, and anticipate and reduce risks.

- Strong communication skills to explain GIS technology to powerful non-experts, work with various personality types easily, and build trust and cooperation across diverse disciplines.
- Management competency regarding personnel, budgets, and schedules.
- The ability to listening to staff at all levels.

If you responded positively to all of the above criteria, you are qualified to fill the roles of champion and manager. If not, be prepared to share the load.

While managers can generally be recruited, champions must arise from within, inspired by a genuine enthusiasm. This makes the champion the most vital individual associated with the project. What if he suddenly leaves? The project will very likely stall, unless it is sufficiently far along that the manager can carry it to completion. Because there are unforeseeable circumstances, grooming a potential successor at the project's outset may be wise. Make sure this "understudy" participates at all important meetings, difficult negotiations, and significant milestones. Get people used to seeing him play an active role. The understudy could be the manager or the person next in line for the champion's administrative position.

Data Conversion Expert

All project staffs include data conversion technicians. People such as survey crew members, data entry specialists, programmers, and photogrammetrists comprise the team that gets the job done. Within this group, there is usually one individual whose technical savvy elevates her to the status of guru. While the technical expert is not officially a member of the management team, she plays too significant a role to be overlooked. The expert may not be an executive decision maker, but her input can dramatically change the course of a project during the planning stages. For this reason, your technical expert must exhibit great integrity. Some individuals may skew facts to promote a personal agenda, and you would be unable to discern the difference—jeopardizing your entire project.

The best technical experts appear professional and communicate well. When a demonstration before the board or potential client is necessary, they put on an impressive show. In addition, when attending committee meetings, they are aware of the protocol in formal environments. This is important, because your committee needs an expert's input, but you do not want a guru who interrupts or belittles the individuals.

Ultimately, you cannot accurately plan, budget, schedule, or evaluate your project sufficiently without technical experts. As such, individuals who understand the complexities of GIS hardware, software, and data should be prized. Make an effort to hire the most talented individuals, and actively develop their careers and personalities. Subscribe to the journals they need, send them to training events and seminars regularly, and keep them abreast of management issues.

Computer Expert

Your project will also require computer expertise. Computer personnel with a background in system administration are critical to data conversion and successful GIS implementation. These people will maintain vital connections among your hardware, software, and network throughout the conversion process. Many system administrators have a background in computer programming and can provide automated methods for creating databases critical to your project. They also may be able to assist with project planning and scheduling by estimating the computer time necessary to run automated image processing and analysis code on the network.

Cartographic and Geographic Expert

Expertise in geography, cartography, and GIS is also critical to the success of your project. Such experts can offer insight regarding database design, software and hardware configuration, and GIS system selection. For example, many users of desktop GIS programs are unaware of the consequences of using incorrect datum and map projections for projects because many systems allow you to display data without defining these characteristics. GIS and cartographic experts can help you avoid these problems.

Many GIS users today are engineers and social scientists, not geographers or cartographers. GIS use has spread to all disciplines because of its usefulness. As a result, many new GIS users begin in desktop environments and without the necessary background in geodetic theory required to design a GIS database. A growing number of universities, witnessing this knowledge gap, have created GIS certificate programs to teach basic skills in spatial referencing, geodetic theory, database design, and spatial analysis. Often taught at night or on weekends, these programs are geared toward busy professionals looking to apply GIS in their work.

Consultant and Conversion Vendor

According to the *Supervisor's Guide to Employment Practices* (Clement Communications, Inc., 1997), in the last three years, there has been a 19 percent increase in outsourcing use. This, along with the trend toward privatization of services traditionally performed by the public sector, means consultants are playing an increasing role in data conversion. Consultants are most often private firms with experience managing GIS projects, although they can also be public entities, such as universities and government agencies, that provide technical and management support, sometimes at no charge. One university in the Midwest employs its students to offer data conversion services, such as digitizing and vectorizing, to nearby municipalities for discounted rates.

A consultant can act as a project champion, manager, technical expert, or all three. The arrangement can exist for the duration of the project or specific phases. The depth of the consultant's authority depends on what is specified in the contract, which should spell out explicitly the consultant's decision making powers. In the end, you cannot cover every contingency, so it is essential that your consultant be trustworthy. This issue is covered in greater detail in a later section of the chapter.

Administrative Support Personnel

Administrative support in a GIS data conversion project is vital. People in administrative support roles can help track the progress of key elements of your conversion project, and they often know how to weed through bureaucratic red tape to secure resource allocation. Even though administrative support staff do not have direct approval authority, they usually know how to "grease the wheels" to shorten time lines for approval.

Digitizer

Because so many conversion projects involve digitizing technology, it is wise to budget for and recruit digitizing personnel. Digitizing tasks are often left to students, temporary employees, or interns learning GIS from the ground floor. Even if you are not planning to use a digitizer with your project (because you opt to scan data, for example), someone with digitizing experience will be an asset to extract attribute information correctly from source data and ensure that quality control procedures are followed.

People familiar with quality control and quality assurance methods are important to a data conversion project. They can take the form of computer experts, GIS users, data developers, or other staff members with a knack for attention to detail.

Hardware and Software

Determining Appropriate Equipment

Hardware and software requirements greatly affect the budgets of data conversion projects. Once you settle on the target system and data model in the planning stage, you can address hardware and software requirements. The first step is to determine what equipment and software is currently available. The software you choose must be the most current version, and your hardware should be readily available and in working order. If your hardware has been temperamental or your software is old, you may encounter trouble. Hardware crashes in the middle of intense data processing could delay the project for days. Outdated software may make it difficult to obtaining technical support or programming assistance. A solution may present itself, but it is wise to rely on the most recent software version. This is particularly helpful if later software corrects bugs found in earlier releases. Determine the software you will rely on, and allocate funds to update and maintain it. Hardware and software maintenance or support can provide sufficient benefit to justify their expense.

For example, one group purchased a scanner for a data development project, but neglected to purchase a support agreement. When the group received the scanner, it was inoperable with the group's current network configuration, and the system administrator was unable to correct the problem. The company that sold the scanner said the 30-day warranty had expired, and requested an additional $2,000 to help set up the equipment. The group was extremely dissatisfied, but unable to resolve the problem on its own. It had no choice but to pay the money. Before purchasing any equipment, be aware of limits regarding support or short-term technical help available from the vendor.

Plotters

Other equipment (and software to run the equipment) also may be valuable. Plotters are essential for data checking in a conversion project. Check plots of the data must be performed to evaluate data shifting, data layer spatial relationships, and attribute errors. Because GIS packages contain built in drivers for the most popular plotters, installing a plotter should be straightforward. You can expect maximum flexibility from plotters that act as a node on the network. In this configuration, they can be accessed by the widest variety of machines.

✓ **TIP:** *Make sure you have all the correct plotter drivers to support your intended plotting formats. Research the formats supported by the software packages you use. The time to purchase and install these drivers is at the start of the project.*

Data Manipulation Requirements

A variety of data manipulation equipment is available for data conversion projects, including digitizers, global positioning systems, digital cameras, and image processing software. Additional software for data manipulation may be required if available data are not compatible with your GIS. Commercial vendors have created conversion software to convert data from unreadable to readable formats. Be aware of the caveats when working with third party software, however. For more information, see the "Guarding Against 'Foolproof' Claims" section in Chapter 4, "Project Planning and Management."

Budget

Budgets for data conversion projects can be quite difficult to determine. Be sure to allocate sufficient funds so you can afford to hire experienced personnel, and try to draw parallels with past conversion efforts or refer to estimates from similar projects. Focus on the benefits high quality data will bring to your organization without getting frightened by the price tag. Because budgets vary by organization and type of agency, one size does not fit all.

Up-front Funding

First, allocate sufficient funding to ensure that the pilot is completed. The pilot should help convince personnel (especially decision makers who control budgets) about the value of GIS data conversion. After all, few organizations will put thousands—or even hundreds of thousands—of dollars toward a project they believe will yield little benefit. Putting enough money up front to safeguard the pilot without a budget crisis is essential. Many organizations put the remaining budget in a contingency fund and allocate money that management deems appropriate after the pilot has confirmed the project's benefits and necessary changes to the project plan are complete.

For example, one organization chose to bring in a data conversion expert to get their project kicked off correctly. This individual was expensive, but able to teach the staff valuable lessons. Not only was the pilot project a success, it was completed ahead of schedule because of the shorter learning curve involved with approaching the project efficiently from the beginning.

Budget Flexibility

Secondly, plan for technical difficulties, and allow some budget flexibility in this regard. Military strategists know that flexibility is critical in situations with many unknown variables. This principle certainly applies to many undertakings, including data conversion. Difficulties can include lost time because necessary equipment is backordered or a loss of electrical power resulting from the unpredictable digging of a backhoe. Equipment will crash, software will have bugs, and staff will need to take unplanned leaves for personal reasons. Be flexible in your budget, and plan well. If you work in a government agency, do not plan to order essential equipment in October if it is the first month of your fiscal year. Government organizations rarely are fully funded at the beginning of the fiscal year, because of budget approvals and other legislation by relevant governing bodies. Nor should you plan to hire a digitizer four months before you are scheduled to receive the proper equipment. The correct order and scheduling of events will save your organization money.

Conversion Structures

Departments

A department may be created specifically to carry out a project, or an existing department may be designated to add a project to its schedule. The difference is significant.

The chief advantage of a newly created department is the ability of the manager and staff to focus solely on the project at hand. That is, after all, the reason the department was formed. When you are going to undertake a complex project (such as GIS data conversion), concentrating your resources can be quite beneficial. Budgeting and scheduling become simpler and you do not have to contend with key personnel or equipment being pulled off your project because of other departmental intrusions.

As appealing as this sounds, there are far more compelling reasons to avoid creating a new department. First, a new department is just that—new. It has no track record. Its personnel probably have not worked together extensively before, so there is little team chemistry. Nor do new departments have relationships with other departments in the organization. They have yet to earn the trust of (or offend) others. Even more importantly, a new department has no chips to cash in, because it has no track record.

Cells

Realizing the struggles a new department may face, but still attracted by the potential of singular focus, some project managers have devised subdepartmental structures called "cells" to coordinate their projects. A cell is given a separate budget independent from the department and its schedule and personnel are dedicated to a given project. Thus, the focus that everyone covets is attained and the disadvantages of new departments are circumvented. The personnel within the cell have already worked together and can draw upon established relationships when interacting with other departments. No new space is required. Finally, failure by a cell is balanced by the successes of the department as a whole. Personnel within poorly functioning cells can be given a fresh start elsewhere, without the attention a full departmental restructuring would attract.

Trying to push a multifaceted project through a department without the use of subunits such as cells is difficult at best, because too much overlap occurs. Either the project intrudes on the department's daily operations (causing neglect and resentment) or it is continually set aside until the deadline nears, at which time a frenzied rush to finish the project will ensue.

Multi-participant GIS

A growing number of conversion projects are undertaken by more than one organization or agency. Multi-participant GIS conversion projects can take several forms.

Consortiums

A consortium is a voluntary agreement among independent entities to cooperatively work together and share resources for the good of all participants. By strict definition, consortium undertakings are beyond the resources of a single member—a significant distinction, because it highlights their dependent nature. No single party has the resources to back out and complete the project alone, so cooperation is easier to achieve. Members are on a relatively equal plane, which makes them more willing to compromise and hear suggestions.

A formal agreement among participants generally covers technical staff, hardware, software, products, and services comprising the project. Each participant must know what its responsibilities are. For example, will the consortium share the cost of GPS receivers, or must you purchase your own?

The agreement you sign must be specific enough to provide a framework, but flexible enough to handle future adjustments, which are sure to arise.

Consortiums have become quite popular in the area of GIS data conversion, primarily because they can divide the high cost of projects among several investors. When digital orthophotography became practical for GIS base maps in the mid-1990s, seven Midwestern counties were brought together by a consultant to share the cost of bringing the technology to the desktops of land conservation and planning departments. All counties were rural, without large budgets or staff to modernize land records. Orthophotography was an efficient method to provide a variety of rural GIS applications while having to purchase only one data product. The consortium made sense because the counties were close together and could share the cost of the aerial flight. Furthermore, photogrammetric control along the border of one county could be used for the adjacent county without additional cost. As it turned out, these counties did indeed derive significant benefits from the project.

For the first time, the counties had real GIS functionality and could perform services for citizens that would have otherwise been impossible. Nevertheless, at least one of the participants admitted that he had been picked to oversee the use of the new GIS despite never having worked in the field. (He was selected simply because he was one of the few county employees who knew how to use a computer.) He grew to appreciate and enjoy the advantages of GIS and today he is in the uncomfortable position of having to perform two jobs at once—his original job and his new GIS duties, which continue to expand. He is lobbying for a full-time GIS position, but it is unlikely his small county can afford to create it. This is a classic illustration of one of the potential weaknesses in consortium projects: the consortium helped its members develop and implement a project they could not have carried out on their own. However, they now must maintain and utilize a sophisticated technology with independent—and limited—resources.

Electronic Cooperatives

One way a consortium can overcome its weaknesses is by expanding its mission beyond a project's boundaries. This kind of organization is becoming more practical and attractive as electronic network infrastructure such as the Internet grows more mature. Several different entities, located thousands of miles apart, can truly participate in real-time data collection, conversion, and analysis, as well as application development, use, and support.

Traditional hierarchical institutions have erected walls to prevent interaction with other organizations and cooperation, but these walls are being undercut by electronic networks. Something about the Internet promotes informal exchanges of ideas—making it a primary building block for future cooperation. New tools for communication (and, by extension, cooperative data conversion) such World Wide Web sites, chat rooms, video conferencing, network file transfers, remote computer log-in, terminal emulation, and e-mail do not require major financial investment or hard-to-obtain equipment. They are readily available with the purchase of today's average desktop computer package.

The electronic cooperative is ideally suited for projects that encompass a large geographic area, such as the National Ice Age Trail digital mapping and database compilation project currently being conducted across the northern United States. Because segments of the trail traverse dozens of counties, coordinators for the project must rely on remote data collection and conversion. Although there is no formal data conversion arrangement, an informal cooperative—a virtual organization—exists nonetheless.

Electronic cooperatives are also a good match for small budget organizations, who may exchange printing on a color plotter, for example, with writing data to a CD-ROM.

More innovative and extensive electronic cooperatives are bound to form in the future, because the possibilities for trimming costs and cutting conversion time are endless. Recent enhancements in two digital data conversion technologies, GPS and satellite imagery, only enhance the possibilities.

Of course, cooperatives are not without challenge, the most pressing of which is probably data security. Before you open your system to outsiders, you must take the necessary steps to protect sensitive data sets.

Partnerships of Essentially Equal Partners

Partnerships in practice may be very similar to consortiums, but the motivations driving participants may be quite different. Consortium members join together because without each other, they could not accomplish a particular project. Organizations forming partnerships could carry out their projects alone, but economic, technical, or other factors compel them to seek cooperation.

Among the most common data conversion partnerships today are those between municipalities and utility companies. Such partnerships are formed with the goal of creating a digital parcel land base to be used by the municipality for assessment and planning, and as an overlay by the utility company for its infrastructure and facilities. When the partners can work out accuracy and licensing issues, the unions appear to work well, saving taxpayers and ratepayers money.

However, because the players are generally able to accomplish conversions independently, power struggles are far more likely than in a consortium. It is difficult to compromise when you want something and you have the means to get it on your own. You have to continually remind yourself of the partnership advantages that led you to sign an agreement in the first place.

Partnerships also add to the complexity of data conversion. Suddenly, you not only have to contend with delays on your end, but delays on the part of other participants. As a result, your own milestones may be delayed, and the effects ripple further. In addition, in today's volatile job market, you never know how long key individuals are going to remain with your partners. Successors will require time to learn the intricacies of the project—if they even care.

If formed wisely, partnerships can be extremely beneficial, as one data conversion vendor and county proved. The county received free scanned and coded parcels for the southern half of the county in exchange for unlicensed copies of the remainder of its digital spatial data. In this way, the county was able to offer something it had developed anyway, and would still possess after the exchange, for an item of tremendous value. The vendor needed to scan the parcels for another project, and by initiating the exchange, it received valuable county data that would have cost several thousand dollars just to reproduce.

Committees

Typically, committees are formed to ensure that the administration of a group conversion project is democratic. They prevent a single person from becoming a dictator, promote input from all participants, and facilitate integration and cooperation because many departments are well represented. Committees are generally sanctioned by an existing entity or governing body, and they can be advisory in nature (offering non-binding recommendations) or carry a measure of authority. Committees also must have a reason to exist. Their purpose can be implied, but they may function better if the purpose is spelled out clearly in a charter. The document also clarifies guidelines regarding how often and how long the committee meets, how meetings are conducted, and how action taken by the committee is approved and carried out.

If you believe forming a committee would benefit your conversion project, limit the size of the committee to about 10 people. If more people want to join, form subcommittees. If the committee becomes too big, quieter members who ought to be contributing get lost in the crowd, and large committees can break down into factions. Some conflict is normal in a group, but it is easier to control in a smaller setting. If approached openly and reasonably, divergent ideas can eventually blend into creative solutions and lead to better projects.

A committee is usually chaired by the project champion or manager and is composed of department heads, an end user or two, and a technical consultant. Do not feel that you have to limit the membership of the committee to company staff or municipal employees. You should, however, settle on a roster and stick to it. One county GIS project committee had a more or less open door attendance policy for upper-level government officials. Originally designed to encourage interest in GIS and make executives feel welcome, the maneuver backfired when a top county executive who rarely attended meetings (and therefore lacked sufficient background to participate fully) showed up for a critical discussion and vote. The regular attendees became reluctant to stir up controversy and voice difficult items that had to be discussed. The result was a flawed plan approved with little dissension.

Data Sharing Guidelines

Partnerships for data sharing present many challenges, but they generally are well worth the effort. Data partners can save money and other resources and provide a sounding board for ideas regarding GIS applications and configurations. The benefits and trade-offs of such an arrangement must be carefully evaluated to determine the best circumstances. Knowing where to find potential partners is the first step in the process.

Where to Look

Federal Geographic Data Committee

One place to start looking for others interested in data in your geographic region is the Federal Geographic Data Committee (FGDC) National Geospatial Data Clearinghouse. The FGDC Web site contains valuable information, including lists of databases housing geospatial data; query capabilities for multi-site queries; examples of partnering efforts; and helpful documentation. Some data sets listed carry fees, while others are free. However, the availability of more convenient formats and projections justify the fees you might incur. For example, some documented data sets might carry charges, but you may receive them in the format and projection you need—a significant real-time savings.

Academic and State Sources

Colleges with GIS programs frequently have information regarding the use of the technology in the region. These institutions also place students in local GIS work environments to gain experience—a good source of inexpensive data conversion labor. Many colleges have GIS related colloquiums, conferences, and symposiums. Local or regional software user groups or chapters of GIS organizations are great places to identify potential partners.

Most states have GIS councils or boards composed of individuals extremely knowledgeable about their geographic regions. Organizations represented in these councils include state, local, and federal governments. Internet list servers such as *gis-l@geoint.com* and *esri-l@esri.com* are valuable sources of information.

Benefits and Challenges of Data Sharing

Benefits

There are many benefits to collaborating on GIS projects with other organizations, and data conversion has one of the largest returns on investment. View collaborative efforts as opportunities to learn from partners and contribute to a common objective. Data sharing and collaborative efforts can change the way agencies work together and help reengineer business processes. Partnering efforts can also, of course, save money.

In one instance, an Army base, Air Force base, and a municipal government were interested in aerial photos for a particular area. Although the participants had different geospatial extents in mind for the area, by pooling resources they were able to award a larger contract to an aerial photo generation firm and receive a volume discount.

Collaboration can also yield less tangible benefits. Increased data availability among cooperating organizations and access to previously unavailable data are extremely valuable. Opportunities to foster working relationships with other experts in the field for idea sharing have long-term benefits. These contacts are useful as sounding boards for new ideas, to identify prospective employees, and anticipate and plan for your organization's future needs.

In some instances, conversions can be completed more efficiently and effectively because collaborations have higher profiles. The skills you acquire and enhance as you plan and execute such a project will likely expand your professional development as well.

Legal Considerations

Perhaps the most crucial element of partnerships is the formal contract you sign. Do not sign anything without the approval of corporation counsel. However, even this safeguard often is insufficient because most corporate attorneys lack legal expertise regarding GIS and data. Have an experienced GIS consultant review the proposed agreement before you sign. Data conversion projects are simply too costly and complex to leave to chance. If an essential item is neglected in a contract, partners are under no obligation to remedy the matter.

Appearing below are some clauses a partnership contract might contain. By no means exhaustive, the list is intended to focus your initial thoughts.

- Maintain all aspects of the purpose and technical methodology of the Project confidential.
- Pledge to take no unusual or unwarranted action which would in any way hinder the successful undertaking of the Project.
- Share gross profits or losses/expenses directly attributable to the Project in the following proportion: ___________ (the partners fill in the blanks).
- Invest Partner One with final and full authority regarding design and technological implementation of the Project.
- If, after examination of this contract for a period not to exceed ten days after the signing of the contract, any partners wish to absolve the partnership described herein, this contract will be considered null and void by all concerned.

Challenges

The "people factor" is the number one difficulty in GIS project partnerships. When two or more agencies or companies work together, coordination issues and difficulties are compounded. A formal committee and formal agreement should direct the effort. Standards should be agreed to and documented that define data specifications, how the partnership will be organized, and each party's responsibilities. Here are some key questions to consider during the planning stage.

Assigning and Scheduling Data Maintenance

Because your data are only as reliable as their most recent update, outline a very specific maintenance plan. How the data will be applied determines the maintenance schedule. Frequency of updates will vary widely by organization and may vary by theme within an organization. Emergency services require daily—if not hourly—updates on highway and maintenance. The same data may be used by environmental groups to monitor the amount of salt accumulation in waterways adjacent to highways, but an annual update will likely suffice for such an application. The agreement must include a maintenance schedule and personnel assignments for making changes. The location and distribution of the "master" data set must also be included in the plan. Many cooperative efforts divide data maintenance by theme or geographic area. As planned maintenance occurs, document what you alter. Remember, data maintenance includes all forms of data—digital, photogrammetric, paper, and others. Make sure your maintenance plan addresses all data sets.

Determining Appropriate Data Accuracy

Accuracy requirements are also derived from specific data applications. One organization may require sub-meter accuracy for a theme, while another party only needs 5m positional accuracy. Determine what is acceptable and preferable for each data theme. However, remember that the better the data, the more costly their collection will be.

NOTE: *If your organization requires data far more accurate than the partnering organization, you must be prepared to pay additional costs related to the higher level of accuracy.*

Data Ownership Issues

First, collaborators must define what all parties mean by "data ownership." Does the owner only distribute the data? Do data management and data liability belong to the owner? You may find this question easiest to address after all other project responsibilities have been determined and assigned.

Setting Budgets and Schedules

The project schedule and budget must be determined before a partnership agreement is signed. Balancing the budget and schedule can be extremely difficult to plan, especially if the parties are not equal contributors. The consequences of one party bowing out of the project or having its budget cut must be considered. Remember to budget for maintenance and data management once the conversion is complete. The partnership does not dissolve with the conversion of the final data layer.

Conversions Between Systems

GIS project partnerships are easiest when the database design and GIS system used are the same or at least compatible. Unfortunately, this rarely occurs. When data design and GIS system are not identical, a format framework must be developed before initiating the pilot project. A file format compatible with all systems involved may be easy or difficult to determine. A group of industry leaders, the Open GIS Foundation, is currently working to reduce the stress involved in transferring data among GIS systems. For more information about Open GIS, see the foundation's Web site, *http://www.opengis.org*. In the interim, using the Spatial Data Transfer Standard (SDTS) may be a suitable workaround. Most commercial GIS software vendors have implemented SDTS into their software. (For more information on SDTS see Chapter 15, "Spatial Data Transfer Standards." Access the Web site at *http://mcmcweb.er.usgs.gov/sdts*, or send e-mail to *sdts@usgs.gov*.)

Because of its independent nature, database design is not handled as easily by industry groups or standards committees. You must develop a database design tailored to your project before you embark on a pilot. The pilot project will test the effectiveness of the design and its transferability. Once the file format and database design exchange criteria are determined, you must address the means for transferring data. Will the participants have FTP capabilities between systems, or should a tape device or CD-ROM writer be used? These issues must be resolved as soon as possible because of procurement issues.

Things to Remember

Cooperative efforts must be managed appropriately in order to be successful. They are most prosperous when the parties have similar data requirements with regard to scale, accuracy, timeliness, geospatial extent, attributes, and other features. These requirements must be addressed at the outset to ensure that your overall goal is reached. Proper planning and efficient execution will lead to a successful data conversion effort.

Conversion Pilot

The project's pilot is a limited test phase where the system's design and procedures are evaluated. Ideally, the project's benefits are confirmed at this point, and adjustments are made before proceeding. The pilot is also beneficial because it exercises the database design, provides a controlled environment to develop and test procedures, evaluates approaches to building the database, and provides data for estimating the effort and cost of full conversion.

NOTE: *If the pilot yields unsatisfactory results, you must be prepared to start over if necessary.*

Limiting the Sample Area

One rule of thumb in planning a pilot is to limit its geospatial extent to 10 to 20 percent of the overall study area. Pick an area that encompasses the greater variety of data types, sources, and themes to ensure the pilot uncovers as many potential problems as possible.

Full project area with extent of the pilot project highlighted.

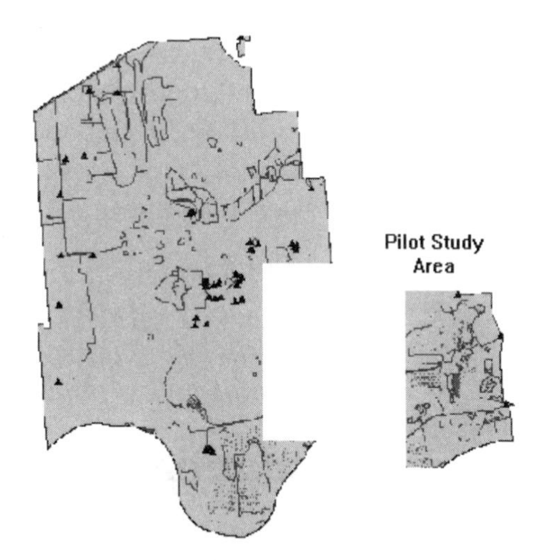

Because the pilot project is a chance to show "non-believers" the value of a GIS data conversion project, it may be smart to pick an area of special interest to dissenters. Organizational priorities also contribute to the area sampled in the pilot project. If the goal of the overall project is to convert data for a large convention or sporting event coming to town, you may want to include traffic, transportation, and parking data related to the event. In short, one effective method of determining the extent of the pilot is to review specifications for the overall project, factor in the political climate, and include information relevant to current and future organizational activities.

Limiting Risk

Maintaining focus is vital for limiting risk in data conversion projects. Well-designed pilots limit risk because of their concentrated focus and trial nature. To further safeguard the project, during the course of the pilot use proven methods when possible regarding format specifications of the target system, and document the methods you use as the pilot progresses. Documentation will reduce the potential for repeating errors or problems you discover during the pilot phase. A detailed record of the time you spend and steps you follow can also be modified into a "how to" guide for the overall conversion process in order to standardize methods used by all staff and decrease the learning curve for new personnel. (Documentation is discussed in detail in Chapter 5, "Documentating a Data Conversion Effort.")

✓ **TIP:** *Remember, the pilot is a test, not "phase one" of the project. However, if it is very successful (i.e., no adjustments are required), then you can consider the pilot a complete first phase.*

Evaluating the Pilot

When the pilot is complete, you have reached the best opportunity to reevaluate the entire project. Design criteria such as the conceptual data model, physical data model, relational database design, data dictionary, and attributes are finessed at this stage. Methods for source preparation and conversion procedures can also be fully appraised. Quality control/quality assurance methods must be evaluated before continuing, and issues regarding database maintenance and backup schedules can be revisited for optimum results.

Your project's schedule, staffing requirements, and budget must also be reevaluated and revised based on the pilot's results. If the pilot identifies significant design problems, be prepared to discard your plan and start over. Keep in mind that it is rare for a pilot to come off without a hitch. Because the purpose of a pilot is to identify problems, do not be discouraged by unacceptable results. Use these results to improve your project.

Conversion Philosophies

There are fundamentally two ways you can tackle a data conversion project: little by little, as executive support, funding, staff, and equipment are available; or conserving resources until you can "blast through" the entire project seemingly overnight. Which way is preferable? The answer depends on many factors, including (but not limited to) your organization's resources, your personality, executive mind set, the nature of a project, and the amount of urgency on the part of those who will use the output.

Incremental Conversion

Technically speaking, all projects are completed in increments, each step building on the previous one. When applied to GIS, this term takes on additional meaning, in the sense that there are planned (or sometimes unplanned) periods of inactivity at one or more points in the life of the project. The interval between activities may be months or years, yielding a project time frame that is more evolutionary than revolutionary.

Implementing a project incrementally breaks it up into bite-sized portions easier for an organization to digest. Everyone involved can push hard to the first project milestone, then step back, catch their breath, and assess progress. The pace makes it less difficult to maintain precision quality control and correct mistakes.

These types of conversion projects also allow you to learn as you go, which is frequently appropriate given the complexity inherent in GIS. You will not be pressured to become an immediate expert on all facets of a particular endeavor. For example, you will not need to know the fine points of cartographic display before you launch a parcel mapping project. Given the extended time frame of the incremental approach, you will have ample time to learn how to create and output your maps before the stage arrives when you have to turn on the plotter.

Incremental conversion is usually found at either end of the organizational spectrum. Large organizations implementing giant projects prefer an incremental approach because anything else would require an enormous mobilization of resources. The National Aerial Photography Program, the digital wetlands inventory, and the digitization of county soils surveys are all classic examples. Certain priority areas are targeted for initial conversion, and then other areas are addressed as time and resources permit. Eventually, after several years, the progress chart is filled in.

Likewise, small organizations choose the incremental approach because their budgets are small and their staff limited. They work on the project when they have the opportunity. A planning department, for example, may digitize zoning maps on days when the weather inhibits field inspections.

Pitfalls

Despite its advantages, incremental conversion is not without potential problems.

Employee Turnover

Employees do not remain with organizations as long as they once did. The longer a project continues, the more likely it is that a key staff member will depart, frequently taking with him irreplaceable knowledge. This is especially troublesome in small organizations. If a planning department loses its zoning administrator, the next one may not be as computer proficient. At that point, the project is effectively dead, because no one with the appropriate skills remains.

Employee Burnout

Even when there is no staff turnover, the project is not safe. Burnout becomes an issue as individuals continue putting time into a project that never seems to finish. Today's technicians have grown up in a world of instantaneous results, giving them less tolerance for slow-paced projects. One way to combat this is to make your milestones bigger events than usual, which creates a sense of accomplishment. Another is to create mini-applications using partial data sets throughout the project, rather than waiting until the entire data set is complete.

Impatience

Fortunately, these are the same tactics you can use to stave off the crippling effects of waning executive enthusiasm and impatient end users. Stockholders put pressure on CEOs and managers, and voters put pressure on municipal officials to show fast results. At any rate, you always have someone looking over your shoulder asking, "When are you going to be done with that?"

Changing Priorities

Growing out of executive and end user concern is the deeper issue of an organization's fundamental needs and goals. Although they are traditionally stable, these needs and goals can change over time. In other words, while it may have been critical to carry out a certain agenda when the project began several years ago, such an agenda may no longer be as important. Simply put, an incremental project can outlast its usefulness.

Changing Technology

Not only can a project be overtaken by organizational change, but projects are increasingly outpaced by technology. Depending on the magazine you read, computer processing speed is doubling roughly every two years. Satellite, network, and GPS technologies are blossoming, with far-reaching implications. Advances in data conversion technology impact not only the form and functionality of an end product, but the entire way projects are structured to meet organizational needs. Commercial cartography made the leap in the late 1980s from manually scribed Mylar maps to computer drafted digital maps. But the impact of CAD went beyond the form of the end product.

If new technology influenced only the surface aspects of a project's output, the impact would be moderate. However, the real power of technology lies in its ability to reengineer entire organizations. Projects that bridge conversion methods may take advantage of the cosmetic enhancements a new breakthrough offers, but they are usually unable to reverse enough to benefit from the deeper possibilities that a fresh technology brings. Incremental projects face the possibility of being initiated by means of one technology, adjusted halfway through by another, then completed by still a third. Such a project will produce output that lacks continuity, compatibility, and full functionality.

Changing Data

Finally, most incremental data conversion requires a static planet to work best. As long as no parcels split, no highways are built, no terrain leveled, no buildings razed, and no rivers change course, there is nothing to worry about. Mapping the moon is an ideal project—the data remain fresh for eons. Here on Earth, however, data have a shelf life. With each passing day, your carefully gathered and formatted data lose value. If you do not use them immediately, they will eventually cease to accurately represent reality, and you will have failed to maximize your investment. Do you remember that quarter section you spent days agonizing over two years ago? A new survey just arrived that alters the position of every feature in the quarter. During the lengthy life span of an incremental data conversion project, this scenario will repeat itself over and over before you put the end product to use. One tactic to counteract this drawback is to develop applications that make use of whatever data you have finished. Nevertheless, the most effective applications are comprehensive. End users lose faith in a patchwork system where parcels are complete in section ten but not in section nine, and intermittent planimetric features are coded.

Blitzkrieg Conversion

Non-incremental conversion is a more scholarly label for the alternative to incremental data conversion, but it fails to capture the flurry of activity this approach entails. Webster's defines "blitzkrieg" as a war conducted with lightning speed by massed air forces and mechanized ground forces in close coordination. Such data conversion is carried out as quickly as possible, using the best equipment and people you can afford in a precisely coordinated manner.

In response to natural disasters such as floods and hurricanes, data conversion projects have been developed and implemented in months or weeks—so why should it take a municipality five years? Nor is it the case that, to complete a project quickly, you must sacrifice accuracy, hire an expensive consultant, and spend huge amounts of money. One 35,000-parcel county went from having no conversion hardware, software, or digital data to a multi-functional, networked GIS with three dozen spatial data layers, including COGOd parcels, in just over two years. What's more, the data are accessible through the World Wide Web. All this was accomplished without any full-time data conversion staff and with minimal outsourcing. How was it done? Through the efforts of top-notch interns, innovative partnerships with the private sector, careful management, and plenty of hard work.

One element that characterizes a blitzkrieg approach is teamwork. No organization can successfully undertake such a project without it. If your organization does not already possess teamwork, you must develop it as you proceed. Because teamwork itself is prized, its development is a serendipitous outgrowth of the blitzkrieg approach. Even long after the project has been completed, you will notice the increased productivity that teamwork fosters, whether in new projects or daily operations. While the blitzkrieg approach is no doubt challenging, it can also be a very positive experience.

Looking for more concrete benefits? The blitzkrieg approach ensures that you attain current organizational goals and needs, because the project is completed before they change. Secondly, executive and end user enthusiasm and support are more likely to remain constant, because both groups can readily see the rapid progress being made. You do not have to sidetrack your project to create mini-applications to maintain everyone's interest and support. They can see that full-blown applications are clearly on the horizon. And, the sheer effort of what you are accomplishing is impressive. Finally, executives can more easily track expenditures to results in a short-term project. Few characteristics are more effective for maintaining current support and securing future funding.

Focus, which is essential to all facets of an organization's success, is sharper when projects are short term. You are forced to fix your eyes on the goals and needs that truly matter. Organizations distracted by the latest technology, funding resources, or special interests find it more difficult to accomplish their primary objectives.

Finally, blitzkrieg data conversion enables you to maximize your investment in the data being converted. Because the data are converted and applied so rapidly, there is little time for them to lose value. The most valuable data are not those that require the greatest resources to convert. Rather, the data integral to the greatest number of applications throughout a project are the most critical. That life span begins the moment each element is converted, not at the end of the project.

Blitzkrieg Drawbacks

On the downside—and every strategy has one—the blitzkrieg approach is demanding on several fronts. At the outset, blitzkrieg projects require a critical mass of equipment, personnel, and money. You cannot wait for budget outlays five years down the road. Consequently, you may have to spend a considerable amount of time before the project can begin waiting for resources to accumulate. Someone might ask, what is the difference between waiting three years for resources, then doing the project in two years and simply starting the project immediately and finishing it incrementally in five years? Well, both organizations finish at the same time, but only one has five-year-old data.

Getting It Right the First Time

Next, blitzkrieg demands a dynamic project champion/manager and highly skilled technicians. There is no grace period for mismanagement and no time for on-the-job training. Your team must be able to do it correctly and fast the first time. How can you judge the capabilities of your team? Sometimes the only way is to throw them into battle. If this is your situation, keep the project as small as you can. Better to test your team when the stakes are not as high.

To pull off this approach, you may need to conscript more troops. The issue then becomes, what do you do with these individuals after the project is complete? Internships are one solution, but interns are frequently inexperienced and unpredictable.

Cutting Corners

Another potential pitfall to blitzkrieg projects involves the temptation to perform substandard work—in other words, cut corners when time is running out. Unless you are doing all the work yourself and you have perfect self control, there is practically no way to eliminate this risk, and it will almost certainly take place, given human nature. The best strategy to counteract this tendency is to insist on accountability. Participants in the project should initial their work, fill out quality control checklists, and be made aware at the start of the project that random auditing of procedures and output will be undertaken. Even then, there are subtle ways to cut corners that can seldom be detected. The effect this will have on a project will naturally vary, but it will never be fully positive.

Team Disintegration

Just as blitzkrieg conversion can build teamwork it can tear an organization apart, especially an unhealthy one. If the intensity of the effort does not bond the participants, it will turn them against each other out of frustration if the project encounters problems. This point alone is enough to send many project managers fleeing back to the relative safety of incremental conversion. However, if your organization is healthy and your team battle hardened you can usually proceed without worry.

Not for the Faint of Heart

Finally, blitzkrieg demands the highest level of commitment from an organization. A half-hearted approach spells doom. Everyone connected with the project must believe that the project is worth doing as rapidly as possible. Even one important naysayer can ruin the best plans.

While neither strategy is perfect, one may suit your organization. You will likely have to adjust a strategy to fit your needs or perhaps create your own approach. Ultimately, you will need an overall strategy to provide a framework for the project and management elements. When you have these elements in place, and they are the best ones for your organization, you will be prepared to carry out a successful data conversion project.

References

Clement Communications, Inc. May 19, 1997. *Supervisor's Guide to Employment Practices.*

Frame, J. Davidson. 1995. *Managing Projects in Organizations.* Jossey-Bass Publications.

National Academy of Public Administration Foundation. 1995. *Information Technology Procurement: Moving from Contracts to Partnerships.* Washington, D.C.

O'Connell, Fergus. 1994. *How To Run Successful Projects.* Prentice Hall.

Acknowledgments

I thank the editor, Pat Hohl, for his patience and assistance, for forming this group of collaborators, and for letting me be a part of the project. I would also like to thank the U.S. Army Construction Engineering Research Laboratories, and especially Bill Goran, for the trust and confidence exhibited in my abilities. Finally, I would like to thank my spouse Chris, son Clay, and the rest of my family, who may have doubted my ability to find a job as a geographer, but supported my chosen career field nonetheless.

Kelly M. Dilks

Project Planning and Management

Kevin Struck and Kelly M. Dilks

The field of GIS data conversion is dominated by professionals trained in geography, engineering, and computer science. Consequently, there is a tendency to view conversion undertakings as data or computer problems, rather than multifaceted projects. The standard approach might go something like this: understand the data fully enough, write sophisticated programs, throw all the computer muscle you can at it, and you will get the job done.

This emphasis on technology and data—at the expense of project management—neglects important details such as cost, leadership, staffing, scheduling, and goal accomplishment. Normally brilliant individuals who understand the value of robust hardware and processing power brush off needs assessment and prioritizing as theoretical extras best left to academics. Regrettably, the results in these instances are less than impressive.

How important is competent project management? According to the U.S. General Accounting Office (GAO), "Developing and modernizing government information systems is a difficult and complex process. Again and again, projects have run into serious trouble, despite hard work by dedicated staff. They are developed late, fail to work as planned, and cost millions—even hundreds of millions—more than expected. The results, in missed benefits and misspent money, can be found throughout government." And these problems are not confined to the government. A *Computer Weekly* article reported that roughly a third of business computer projects are "runaways"—behind schedule, over budget, and unlikely to meet key objectives.

Successful data conversion does not just happen. It must be structured to promote systematic planning, monitor work, and measure results to ensure that you obtain what you set out to accomplish.

Project Elements

The structure best suited to data conversion is called a project. The definition of a project shows how well it fits data conversion. Every project has the following characteristics.

- Single, unique focus (goal oriented)
- Start and finish
- Limited resources
- Involvement of an ad hoc, cross-functional group of people
- Sequence of coordinated, interdependent activities
- Specific end result
- Definite user (i.e., client or customer) who will use the results

How does data conversion fit into the project framework outlined above? Take parcel mapping, for example. You have been asked to create a digital parcel map for your city. Because no one has completed this task before, it is certainly unique, with a special set of problems to identify and overcome. Although there will always be updates, the initial conversion is a onetime event, with a definite start and end, at which time the map becomes functional for intended end users. Even though you might ache for unlimited time and money, you will likely find yourself facing very limited resources. You may need the help of surveyors, assessors, computer experts, and accountants, and you will have to coordinate their activities. When you are finished, a digital parcel map that meets your specifications will be complete, and users will apply the new parcel map for a variety of purposes.

This brief example illustrates that data conversion involves more than simply sitting down at the best computer you can find and cranking out parcels. *You ignore project management at your own peril.* Without it, you will waste time and money, alienate staff, upset executives, disappoint end users, and frustrate yourself.

A successful data conversion project consists of the steps below.

1. Planning
2. Articulating the plan
3. Gathering resources
4. Working toward milestones
5. Demonstrating usability

It is helpful to break down the steps into practical smaller steps you can use as a ready guide. This discussion assumes that you already have been given a general directive to carry out. An executive, committee, or client has come to you with an assignment such as modernizing assessment records, vectorizing Mylar zoning overlays, or converting hardcopy plats to electronic images.

If the directive is too broad, first separate it into multiple projects. Remember that a project is defined as having a single, unique focus. Vectorizing Mylar zoning overlays is already a single focus, as is converting hardcopy plats to images. However, modernizing assessment records describes a series of projects involving maps, tabular records, and images. It is impossible to pinpoint a single task that, if completed, would fulfill the directive. If you have difficulty applying the five steps above, then the assignment is probably too broad. Break it down into projects that can be addressed with a single task, no matter how large. Otherwise, you will find completing the project maddening.

Planning

Begin by comparing the directive you have been given to similar efforts to get a feel for what it will require. Take a quick tour of an organization that has completed a similar project. Keep the tour short—you only want an overview. Avoid returning home with a ready-made formula for how to tackle your project because—remember—your project is unique. Your end users are different, and so are the visions of your executives. It is too early to allow biases to form in your mind. Maintain an open mind as you read case studies in GIS related magazines and talk to peers. If you have already become convinced that scanning parcel maps is more prudent than using coordinate geometry, you will fail to hear end users when they explain the parcels will constitute a base for utility mapping.

Obtain Input and Support from Executives

Executives within your organization may have already met and settled a number of issues regarding the project before you became involved. They may have prepared a goal statement, measurable outcomes, time line, and budget.

Be flexible with this group. If you are invited to run the meeting, all the better. Otherwise, defer to the project originator and assume the role of consultant. If you are assigned the task of drafting a "guest list" of meeting attendees, be certain to invite every executive remotely affected. Avoid insulting people or giving individuals reasons to resist the project.

During the meeting, take thorough notes. Document the expectations and concerns of the group to guide your planning efforts and protect yourself. Make note of issues the group does not appear to have strong opinions about—again, for your own protection. In addition, be careful not to create unreasonable expectations, because the group is likely to latch onto ideas that may not be within the scope of the project. Working with such a group is a bit like waiting for the handoff from the quarterback: once you get the ball, you will be expected to run with it. However, do not take the ball prematurely, and once you get it keep it. A strong, knowledgeable group of executives may have difficulty watching you control the project. Make it clear that they will be fully informed of the project as it progresses, but you must have autonomy. If they disagree, suggest stepping down, because the project may be doomed from the start.

Of equal concern are executives who know little about your project beyond a vague directive. One county executive put it this way: "All the other counties around here are doing great things with GIS—why aren't we?" They have expectations, but they cannot understand the details or implications of their expectations. Rest assured, as the project proceeds, their opinions will likely become more frequent and firm.

To avoid surprises, educate executives from the outset so they can express their expectations as early as possible. Explain what surrounding municipalities have done and why, and explain what would happen—good or bad—if the same approaches are instituted locally. Make the presentation as visual as possible, and distribute short summaries. Allow time for information to be absorbed, and then reconvene the group for feedback.

Perform Needs Assessments of End Users

End users know better than anyone else what functionality they require and what they wish they could do. This is the primary input you should collect from them. You may already have been told by your executives that municipal clerks and planners need digital parcel mapping, but you must learn why. There are a variety of ways to convert hard-copy maps to digital form, each of which results in a different end product capable of supporting various applications. Secondly, there are unlimited ways to structure the database linked to the digital map. What data fields should be included, and does it matter whether a data type is defined as a numeral or character? Next, a number of map labeling options exist. Are lot line dimensions assets or liabilities? Finally, how should the mapping be integrated into an overall land information system? This involves issues concerning the operating system, application software, network, and maintenance, to name a few. Without knowing in detail the needs of your end users, you cannot guarantee that the methods and designs you choose will address their daily job requirements.

Standard needs assessment exercises begin with inviting end users to a demonstration of general GIS capabilities to familiarize them with the technology. Individuals attending the meeting are handed questionnaires regarding their job responsibilities. The packet includes a self-addressed stamped envelope that attendees are expected to return within one month. After the questionnaires have been returned, personal interviews are scheduled with the respondents who provided the most detail in their answers. The results of the interviews are compiled into a formal needs assessment, which describes what the end users do and what they perceive to be their most pressing needs.

This method is sound, but it could be improved. Rather than scheduling interviews with only the most articulate respondents, make a point to interview some of the less interested, less vocal users. Find out why they are less enthusiastic or knowledgeable than their peers. Their feedback may uncover another core problem or identify other neglected issues.

Also consider including individuals or organizations who regularly deal with your group of end users. For example, if you are interviewing the staff of the land planning department, interview major customers. Ask a surveyor, septic system installer, and building contractor how well the department meets their needs for information access and processing. When compiling a needs assessment, there is often a tendency to stop at the point where various needs have been identified. A more valuable also evaluates them.

You may survey end users to determine how certain tasks could be performed more efficiently. However, do not expect much useful feedback. While end users are experts at what they do and need, they seldom have the background in GIS to provide specifications for data conversion and application development. Finally, distribute the complete needs assessment report to end users to confirm your findings, and demand feedback.

Establish Goals, Objectives, and Strategies

After gathering input, you are ready to set goals, develop objectives to fulfill those goals, and describe necessary strategies. You have already gathered input from two very important groups of people, executives and end users. Next, you must incorporate opinions from individuals who will work on the project. Their involvement begins when objectives and strategies are developed. Balancing the input from these three groups is a delicate but essential trick.

Effective goal setting usually involves picturing an end result. Good questions to ask are, "How will we know when we are done?" and "What will give us that satisfying sense of achievement?" More concretely, goals are specific, measurable, agreed upon, realistic, and framed in time and/or cost.

In regard to parcel mapping, your goal might be, "Within one year, provide one-screen service for citizens searching for parcel related data." Does this meet the above criteria? You will know you are done when a citizen walks in, brings up a parcel on the public access terminal, and uses the accompanying menus to read ownership, assessment, zoning, and deed information. This statement qualifies as a goal because it is specific (parcel related data for citizens), measurable (one screen, not two), agreed upon, realistic (many municipalities have already achieved this), and framed in time and/or cost (within one year).

Be Specific

Objectives expand the goal into a set of specific steps. "Any objective lacks clarity if, when shown to five people, it is interpreted in multiple ways," writes J. Davidson Frame in *Managing Projects in Organizations* (Jossey-Bass Publications, 1995). Clarity is tantamount when formulating objectives, because it will be difficult to develop strategies if you are not absolutely clear about the objective. Objectives must mean the same thing to you as they do to technical wizards or potential consultants. It may be wise to involve technical staff when forming project objectives. A clear objective for the above example would be, "Create a pulldown menu using Avenue [the ArcView GIS programming language] that will allow the user to display zoning codes for the highlighted parcel directly on the map." In short, strategies must address activities, resources, required skills, time line, budget, assumptions, risks, and contingencies.

One-sentence goals and objectives are ideal, but a one-sentence strategy almost always neglects vital elements. Strategies, like instruction manuals, should describe in detail how a particular objective will be fulfilled. Do not be afraid to state the obvious. People who read, act on, and fulfill your strategies should have no significant questions regarding particular steps.

A strategy for the public access terminal example might read as follows.

- A programmer proficient in the Avenue programming language will write a script that reads the parcel ID of the selected parcel and stores the ID as a variable.
- The script then reads the zoning table, which also contains the parcel ID, locates the corresponding record, reads the zoning code, and labels the selected parcel with the code in the lower center of the parcel.
- Use capital letters, bold, and the color red.
- This project will require a computer equipped with ArcView, Avenue, and sample parcel data, and it should be as similar as possible to future access terminals.
- The project assumes that all parcels have been assigned a single zoning code. However, it is possible that some parcels have not been assigned a code, or that some parcels have multiple zoning codes; all zoning tables must be searched for null values (refer to planning department for consideration).
- The project must adjust the script so parcels with multiple zoning receive multiple labels.
- Time: 3 hours. Cost: $90. Contingency: 2 hours at $60.

Test Project Feasibility

Testing project feasibility is often overlooked. Perhaps because the two biggest hurdles are executive and financial support, when these items are in place project managers take completion for granted. However, failing to analyze other elements may pave the way for change orders, cost overruns, and delays.

When evaluating project feasibility, begin by taking an inventory of resources to see how they match project requirements. Second, conduct a trial run or pilot to test the project's design and your ability to meet its requirements. (See Chapter 3, "Conversion Resources and Structures.") While it may be unpleasant to question technical team skills, the hardware, software, budget, time line, or integrity of source data, this is precisely what must occur.

- **Technical team.** Compile a detailed list of each team member's skills. Consider the individual competent in a certain area if he can carry out the task without supervision. Next, give the person a character rating of "yes" or "no" based on whether he has been an integral part of a successful project in the past. When the time comes for team members to pick up the project and carry it that final half-mile, the only qualities that matter are (1) can they do what needs to be done and (2) can they do it together? While there will be opportunities for novices, avoid including them in your inventory.
- **Hardware capacity.** Do not compromise regarding hardware. Evaluating hardware is meticulous and labor-intensive. Hardware comes with numerous specifications, all of which can hurt the project if overlooked.
- **Networking considerations.** Physical network constraints must also be assessed, especially in data partnerships. Review network connectivity among machines, plotters, printers, digitizers, scanners, and other equipment.
- **Software functionality.** Because all GIS software undergo frequent upgrades, you must evaluate backward and forward compatibility, and the software's harmony with changing operating systems and hardware platforms.
- **Budget.** Overbudgeting and underbudgeting can both impact feasibility. Strike a balance between providing a cushion and maintaining fiscal prudence.

- **Time line.** Apply the same principle here as you would for budgeting, substituting weeks and months for dollars and cents.
- **Source data integrity.** Data of suspect quality sink many data conversion projects. Many map creators are not trained in cartography and do not follow strict guidelines, such as those used by the U.S. Geological Survey's (USGS) map production section. Source data may not contain detailed metadata. These data can be incorporated into the GIS, but keep their accuracy in mind when performing spatial analyses. Diplomatically question data vendors regarding the quality of their data and collection procedures to gain insight into data strength and identify or anticipate problems.
- **Geographical constraints.** Verify that data acquisition is physically possible. Perhaps you should include catch basins as part of your planimetric infrastructure mapping, which can be problematic on crowded city streets. Aerial photos may be available for some sections of your project area, or the scale may be different for each photo transect.
- **Converting sensitive data.** Many agencies take political considerations into account when identifying data to convert first. If possible convert the most politically expedient data first. If you believe certain communities may block data acquisition efforts, assess such resistance at the outset. Sensitive data such as police records or personal income may require special permission or require additional database security measures.

Remember, if you come up short in just one of the areas described above, your project fails feasibility and must be adjusted accordingly.

Prioritize

Prioritizing means ranking the importance of each element within a project. Cost and time overruns may force you to eliminate one or more parts, and losing some financial support has the same effect as a being asked to finish the project two months ahead of schedule. To anticipate such situations, prepare a chart ranking the importance of each project element, so you can make informed decisions if necessary. Base the chart on executive and end user input and priorities that surface as the process moves forward.

Prepare a Budget

See Chapter 3, "Conversion Resources and Structures," for a discussion of budgeting.

Prepare a Time Line

Your time line should reflect the priorities of your organization and be flexible enough to compensate for system crashes, personnel turnover, budget reductions, and emergency situations. Also keep in mind that it can be costly to schedule technology based projects for a life cycle longer than 18 months, or the time it takes computer processing power to double.

A county in eastern Wisconsin has worked on two long range projects simultaneously. The first involves parcel mapping in house over a four-year period and the other involves acquiring planimetric features in a vector format over a two-year period. When the projects were initiated, neither GPS nor satellite imagery were viable conversion tools. Just two years later, both projects would be structured and budgeted far differently to incorporate these revolutionary tools. Currently, such a restructuring is impractical, if not impossible, so the second half of each project will be completed using less than optimal strategies.

Articulating the Plan

Communication is essential for successful project management. Individuals with an interest in a project become offended when they learn of changes or initiatives after the fact. And, the greater their interest, the greater the offense. Several tools to publicize progress appear below.

- Flow diagrams are a simple but effective means to illustrate project steps and uncover missing ones.
- Time lines provide context and show the duration of each phase, enabling colleagues to spot periods where resources may be overextended.
- Spreadsheets are an increasingly essential tool, particularly for budgeting.
- Maps and drawings depict topics quickly and simply.
- Classic pie charts and bar graphs can also be helpful, if used properly and kept simple.

Periodic meetings and newsletters can also keep people posted on the project. The meetings can be brief and informal and the newsletter functional. Send the newsletter at regular intervals to maintain interest. Once or twice during the project, schedule a tour of the work site. Have a sample of the finished product available to give people a personal look at the processes involved.

Gathering Resources

Once the project is planned and articulated, you are ready to gather essential project resources, such as equipment, software, infrastructure, staff, and space. See Chapter 3, "Conversion Resources and Structures," for a discussion of this phase.

Working Toward Milestones

Project milestones mirror project objectives. While you may consider an operational scanner important, it does not qualify as a milestone if the objective of the project is to convert hardcopy data to digital form. Do not mirror goals (they are too general) or strategies (they are methods only). Concentrate on verifying objectives, because they grow out of end user and executive input.

Milestones are defined as specific accomplishments falling within the duration of the project that are easily identified, occur at logical intervals, and fit into a recognizable context.

They determine the health of the project by signaling whether objectives, the budget, schedule, and data quality requirements are being met. Also be aware that not every accomplishment makes an effective milestone. For example, if you are scanning and archiving subdivision plats, an appropriate milestone would be scanning 25 or 30 percent of plats in the project.

Like buoys, milestones can serve as warning markers. Paying attention to them early in the project, while adjustments can still be made, may mean the difference between success and failure. For better or worse, milestones also affect morale. Publicize good news, and resist the impulse to whitewash poor results.

Demonstrating Usability

Project completion represents the ultimate milestone. At this point, many project managers skimp on total project evaluation. Few people want to probe mistakes in their work and summarize them in a report, and others may be equally uncomfortable publicizing successes. Both rationales, while understandable, are unfortunate, because it robs individuals involved with the project of formal performance evaluations and stifles input from potentially dissatisfied end users. Groups planning similar projects are also deprived of your evaluations. Because project evaluations require time and money, incorporate this phase into the project schedule and budget.

There is an increasing trend toward formal project evaluation. Many federal and state grant programs now make project evaluation integral to applications, which must outline a thorough evaluation strategy, identify team members, and describe methods of disseminating findings.

Approximately half of your project evaluation can take place immediately (after you have spent a week in the Caribbean, that is). Obviously, you will know right away whether the project was completed on time, within budget, and according to specifications. You can use a simple tool, called a Gantt chart, to help you and others visualize the project's performance in the first two areas. The chart compares schedule and budget projections against results. A spec sheet or checklist will help you in the third area.

Example of a Gantt chart for project planning.

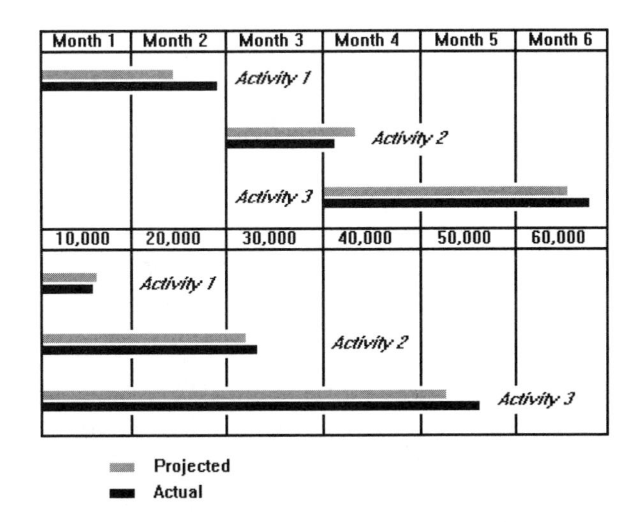

Time, money, and specifications are concrete and unbiased ways to determine project success or failure. More subjective methods involve asking subtle questions, such as are executives satisfied with the outcome? Did the products or services live up to end user expectations? Answering such questions thoroughly takes time and a formal project evaluation composed of the following three parts.

Evaluation Strategy

Identify specific outcomes to measure apart from the big three (time, money, and specs). These outcomes will generally be unique to the project and focus on how end users now perform relevant functions more efficiently. For example, if you supplied assessors or appraisers with digital soils mapping, how does this advancement improve the process? Pinpoint evidence that the improvements have occurred, such as comparing paper based acreage calculations of soils types (and, hence, crop yield projections and land value) made in previous years with calculations using digital tools. Then, document the improvement (or lack thereof).

Team Members

Choose approximately six qualified individuals involved with different aspects of the project. One county applying for a federal grant to put GIS data and functionality onto the desktops of rural clerks and planners assembled a group that consisted of a rural clerk, a county planner, a network specialist, a GIS specialist, and a land records professional from the private sector. Only the individual from the private sector was not a direct project participant.

Disseminating Findings

What mistakes were made? What were the positive impacts? How could the project have been improved? What would you have done differently? What surprises occurred? What facets of the project worked better than expected? What advice can you provide? Address such questions through newsletters, journals, books, user group meetings, the Internet, videos, and conference seminars, to name a few. The core of your presentation should be your evaluation team's findings uncovered in the field. Readers and listeners benefit most from practical case studies.

Some thoughts regarding the use of the selected informational media follow.

- Newsletters can be attractive and low-cost means to disseminate information to a wide audience.
- Local newspapers may prove useful as well. If you submit information, verify that articles are accurate before they go to press.
- GIS periodicals and book publishers are always looking for interesting case studies. In addition, company sponsored publications, such as ESRI's *ArcNews,* can also publish project evaluations in an attractive format.
- Data conversion vendors and public agencies have launched a variety of formats to publicize projects via the Internet. Chat rooms enable interactive discussion, and

e-mail list forums provide daily opportunities for you to describe your application and field questions. Dedicating a Web site to the project—if the project is in fact a Web site—is effective as well.

- The increasing availability of video cameras and digital editing software has fostered the development of local professional video production.
- User group meetings can be useful for test the rough draft of a seminar or conference presentation, gain support, and share your experience.
- Conference and seminar presentations are the most stressful method of publicizing your project, but also the most rewarding. Endure the panic to describe your project to a large audience of peers.

Ultimately, nothing beats showing the executive committee how your products and/or services improve the daily work of end users. On a field trip, select an articulate end user to present evidence that the project is making a difference.

In addition to using an evaluation team, follow the strategy of one project manager who encouraged end users to maintain brief work logs describing tasks they were able to complete more effectively or efficiently because of the products or services the project helped to put in place. Executive committees, clients, and taxpayers need to see how much "bang" they are getting for their buck. Not only does this make them happy, it is their right.

Long-term Data Maintenance

Long-term data maintenance is vital to successful GIS implementation, and maintenance plans must be developed as part of the data conversion plan. Maintaining and managing a data set secures its integrity. Ensure that you set a schedule that includes updating data, performing and storing backups, and safeguarding original data sets. In addition, keep in mind that not all users of the data should have equal access. Save "write" permissions (or the ability to edit) for knowledgable individuals who will work to ensure data integrity.

Strategic Issues

This section discusses guidelines for evaluating whether to conduct a data conversion effort in house or hire a consultant. Criteria for evaluating consultants are offered, along with a discussion of procurement options.

Consultant versus In House

Whether to hire a consultant or conduct a conversion effort internally depends on your organization's situation and the type of project. Three major factors affecting the decision are cost/schedule, quality, and long-term ramifications.

Cost/Schedule

Businesses and agencies operate in different worlds, and the factors that affect them naturally impact costs. A consultant's estimates must include overhead and a margin for profit. Staffing may also be a relevant issue—the agency that has these things in place already can usually beat a consultant on price. Time and desire also bear on the discussion. You may be able to implement a project for half the cost of a consultant's quote, but if you do not have the time or desire cost is meaningless. These variables constitute the consultant's greatest advantage, because all consultants have the desire and they will always find the time.

Consultants may level the playing field in other ways. They usually have more experience estimating costs, because they do so almost every day. This means less over- and underbudgeting. They also have been sharpened by competition, so they may have created innovative (and less expensive) ways to solve problems. Finally, the public sector is frequently guilty of allowing work to expand to meet a seemingly bottomless budget and endless time frame.

Quality

If you assume the in-house and consultant costs are equal, the relevant issue becomes quality. To address quality, define it in terms important to you (e.g., data integrity, freshness, system compatibility, database query speed, cartographic appearance, image resolution, coordinate precision, and so forth).

In general, consultants excel in the areas of quality control and promoting industry standards, while in-house conversion staffs may serve as better project managers and ultimately be more conscientious—if only because they must live with the results of their work. In other areas, such as organizational understanding, access to materials, and equipment, either approach can be equally attractive.

Long-term Ramifications

In the short term, your choice will impact project cost and quality. As time goes by, you will notice other effects, some positive and some negative.

If you hire a consultant, and everything works out as hoped, you will see a synergy (or teamwork) develop that leads to more effective solutions. When this kind of relationship is interdependent—that is, each party voluntarily needs the other—interaction between the two is healthy and mutually beneficial. However, when the relationship becomes dependent to the point where an organization cannot accomplish essential tasks without a consultant's aid, the door is opened for financial abuse. If your choice would make a consultant indispensable to your daily operations, you might want to reconsider.

Choosing to remain independent prevents such abuse. Your staff will be forced to gain their own expertise, but independence cuts you off from the synergistic process. In addition, if you hire staff solely to implement the project, you must find other roles for individuals to fill when it is complete.

Choosing a Consultant

If you have decided to contract out your data conversion, a number of factors can help you narrow the field. The level of experience, service options, and pricing varies for each type of consultant. To locate potential candidates, contact local governmental agencies in your area, peruse GIS magazines and attend conferences, and ask colleagues.

Evaluating Candidates

Because GIS is a holistic blend of disciplines, technologies, and management issues, look for consultants who see the work they propose to do for you as a part of a larger interdependent system, not an isolated project. To help your evaluation, ask for the following items on your tour of a candidate's office.

- Documentation that certified training has been completed for software packages the consultant claims to be using. Inquire about software versions and hardware.
- Resumes of relevant personnel. Some consultants are notorious for hiring less than qualified technicians and (paying them below the industry standard) in order to keep labor costs low.
- Names and phone numbers of the past year's clients (all of them). This is more indicative than a list of hand picked references.
- Step-by-step procedures regarding how quality control is carried out.
- A guarantee that problems will be fixed free of charge. Make note of time limitations and other relevant details, especially on large projects. It may be some time before you can thoroughly analyze all the deliverables.
- An organizational chart, so you can verify supervision of second and third shift crews.
- Summaries of projects similar to yours, including project specifications and planning documents; how the consultant's work was integrated into the client's overall land

information system; personnel and duration of the project; hardware and software used; budget and schedule report (i.e., did change orders exceed 5 percent of total cost?); and project evaluation.

If a consultant gives you an odd look when you ask for formal project summaries, it may be an indication that an out of sight out of mind mentality exists. Look instead for consultants who take the time to formally evaluate and document their projects. This indicates a desire to analyze, learn, and improve and the kind of conscientious consultant well worth your money.

Once you have chosen a firm or individual, you will build a relationship around the project. The consultant needs you, you need the consultant, and the project needs you both. Here are several ways to create synergy and keep the peace.

- Develop project objectives and strategies together. The consultant gains an understanding of the project's requirements, and you gain expertise. This approach also fosters team identity and leads to innovative solutions.
- Do not push a consultant to do everything your way. Keep an open mind. The consultant has dealt with dozens of conversion projects and will probably be able to suggest improvements to your plan.
- Monitor progress in a systematic way. Agree to project milestones and be diligent about checking them. However, be flexible—do not be a slave to deadlines or the scope of work if a change to either might result in a more beneficial implementation. The consultant has not perfected time lines any more than you have.
- Limit change orders. Many consultants complain (and rightly so) about clients who fail to determine what they want. One day they prefer "Street" spelled out and the next day they want all names abbreviated. If you sit down with the consultant prior to the project to discuss

exactly what product you need and why, and if you keep the project's time line short to avoid obsolescence, change orders should be rare.

- Check deliverables as soon as possible so problems can be addressed. This is the ideal, but it is not always practical. To save time, randomly check areas you suspect may have caused trouble. If you find errors, keep a level head. Good consultants will trip over themselves to make things right. Agree on a reasonable turnaround time to check deliveries.
- Focus on solving problems rather than assigning blame. If you feel you really must find someone to blame, wait until you have solved the problem.

Traditional Procurement

If you plan to use a consultant, you will inevitably face proposal and contract issues. Currently, government and industry are striving to find less time-consuming, more effective, and less adversarial ways to procure data conversion technology and products. The California Task Force on Government Technology Policy and Procurement report found that "The policies and procedures instituted to ensure that IT [information technology] expenditures are appropriate have created an environment in which it takes too long to develop an IT solution from conception to implementation; problems or mistakes are not quickly surfaced; projects are subject to delays and cost overruns; more appropriate technologies are often bypassed in favor of an outdated solution; and an adversarial relationship between the State and its vendors prevails." Although this does not address data conversion projects per se, the procurement of information technology is similar.

Traditional competitive bidding practices, or Request for Proposals (RFPs), have incorporated the two-envelope system, the prequalification method, or a combination of both. Having determined your organization's needs, usually through a needs assessment and requirements analysis, you assess ways to meet those needs. Once you settle on the best solution, you draft an RFP, open the first set of envelopes containing the proposals of the respondents, evaluate them on the basis of some type of weighted scoring mechanism, then open the previously guarded second set of envelopes containing price quotes.

While there are several junctures where this process can go awry, the most crucial step is determining and describing your data conversion needs. If you are not absolutely certain what you need, or if you are unable to clearly specify your needs to others, the responses and quotes you receive are bound to be unsatisfactory. If this manifests itself during the RFP evaluation stage, the result is one or more bidders who disqualify themselves by responding inappropriately to your request. "Well, that's their tough luck," you might say. However, it is your tough luck, too, because these bidders might have given you the best service and value if they had been able to determine what you were requesting. Furthermore, if misunderstandings regarding project specifications occur after the winning consultant has been selected, you really have problems. Things can become quite messy when six months into the project you explain to the consultant that you thought it was clear you needed attribute coding and the consultant responds by sending you a bill for costs beyond the original bid.

Parts of a Data Conversion RFP

A. Contract information

1. Instructions for labeling and preparing proposal materials
2. Number of copies
3. Address
4. Expiration date and time
5. Bond percentage
6. Contact person

B. Proposal outline

1. Section headings
2. Synopsis of each section

C. Executive summary

1. Background on your organization
2. The fundamental nature of the proposal in non-technical language
3. Description of the general benefits expected

D. Organizational inventory or environment

1. Description of your current and/or future digital infrastructure
2. Description of your current and/or future land records data
3. Description of the data product's end users

E. Statement of work for the project

1. Nature of the problem
2. End product specifications
3. Strategy to solve the problem and arrive at desired end product
4. Integration of the end product into your existing and/or future systems
5. Metadata requirements
6. Pilot project specifications, if necessary
7. Technical support and training, if necessary

F. Consultant qualifications

1. Size, years in business, and financial status
2. References
3. Number of similar projects completed, and five most recent summaries

4. Field support logistics
5. Subcontractor information

G. Proposal instructions for the consultant

1. Provide information related to firm qualifications
2. Describe in detail project implementation procedures, schedules, and so forth
3. Explain how the needs of the organization and end users will be met
4. Describe strategies for integrating the data into the existing system
5. Provide any additional information you feel is beneficial

H. Selection criteria

1. Format (committee, panel, third party, individual, etc.)
2. Faithfulness of responses to RFP
3. Equipment, software, and/or product demonstrations
4. Data integrity guarantees
5. Reasonableness of cost and time estimates
6. Qualifications

I. Miscellaneous

1. Procedure for changes in RFP
2. Waivers, disclaimers, and liabilities
3. Public record requirements

Competitive bidding seems to work best for generic or simple projects. The more extensive and complex the project is, the more difficult it becomes for a consultant to formulate an accurate bid. There are other problems with the bidding process. Some consultants, desperate for business, may low-ball their RFP bid. After the firm is awarded the job, however, they may plead unforeseen circumstances to justify additional charges. A few consultants are able to faithfully stick to unusually low bids, but they have to skimp on staff, training, equipment, and quality to succeed. Other times the low bidder is the consultant who has made the most assumptions about the project's scope, not fully understanding the statement of work. On the other hand, consultants may understand the scope of the project better than you do, to the point of seeing gaps and weaknesses in the statement of work—but they will hesitate to address anything that is sure to add to the project cost, and by extension, their bid. Consultants who low-bid may cut corners, ask for money later in the project, or deliver poor quality data. Yet another surprising concern is *high* bidding. One busy consultant was heard saying that he bid a recent proposal extra high, because the only way he was going to do the project was if he got paid extremely well. Guess who was awarded the job anyway?

New Procurement Options

Unsatisfied with traditional procurement results, some agencies have developed or borrowed new approaches. Because the new methods have only been in use a relatively short while and by comparatively few agencies, there is little extensive research to deny or confirm their effectiveness. However, early returns appear favorable.

Performance Based Contracting

As noted previously, one of the major weaknesses of traditional procurement is the assumption respondents must make about the organization's or agency's needs. No matter how well the RFP authors clarify project objectives and strategies, they cannot spell out every detail. In addition, more often than not, the authors have only a general idea of the conversion steps involved. Consequently, their RFPs are easily misinterpreted. One small county had difficulty composing its RFP for just this reason. Rather than bluff their way through the document, county personnel had the humility to admit they were not sure exactly how the job should be tackled. They hired a consultant and the RFP was written more competently. Relying on an outside party to write an RFP has drawbacks, however. You are spending money before you even obtain bids, and vendors who receive your RFP are getting secondhand information. To reduce the possibility for error, explain the job to the consultant, and then listen and ready carefully as the consultant translates your wishes into an RFP.

In performance based contracting (PBC), the statement of work traditionally included in an RFP is eliminated. In PBC the goal is not to articulate conversion steps (which is too complex), but to describe the project's *purpose*—the project's outcome, not the method for reaching it. This leaves methods to respondents, who are free to propose their own procedures. In PBC, the more precise your RFP, the better. Consequently, you limit the RFP to items that are precise, concrete, and quantifiable. A description of what must be done will meet project standards is far better than a description of how the standards must be met. This philosophy can also apply to performance and quality requirements. When you are done your RFP should follow the outline for a traditional RFP, with the items below replacing those listed under "E. Statement of work for the project."

1. Quantifiable performance benchmarks you expect the vendor to meet (e.g., project duration and cost).
2. Quantifiable quality standards you expect the vendor to meet (e.g., positional accuracy and image resolution).

Vendors able to propose solutions that beat your benchmarks and standards should be rewarded accordingly. According to some project guidelines, the savings created should be split evenly between the organization issuing the RFP and the vendor. The U.S. Department of Housing and Urban Development (HUD) has saved more than $30 million since 1991 under a 12-year PBC contract with HUD Integrated Information Processing Services.

California Performance Based Procurement

Editor's note: The following material can also be found in an article titled "Information Technology Procurement: Moving from Contracts to Partnerships," at the following Internet Web address: *http://www.clearlake.ibm.com/Alliance/clusters/crt/proadmaj.nah.html.*

This approach seeks to form a strategic partnership with qualified vendors for the purpose of long-term, mutually beneficial business relations based on trust, honest and open communication, and teamwork. After the business problem has been stated publicly, interested consultants work with your organization to establish their credentials. You competitively review consultant credentials and select a limited pool of qualified candidates who become, in effect, your business partners. Next, your staff work with each of the business partners to fully investigate the existing environment and business problem, and to develop the "best of breed." Put another way, you work with each business partner to make sure their proposal is as good as it can possibly be. Once all conceptual solutions are proposed, a selection using "best value" techniques is made, and contract negotiations begin with the successful consultant.

The contract is for delivery of a working solution, so you accept it only when the project is successful. Payments are provided when, and if, the solution delivers sufficient benefits to cover the cost. This creates a greater sense of urgency on the part of the business partner to implement a solution because payments will not occur until after the solution is implemented and providing benefits.

The shared risk factor is the critical factor in establishing the strategic partnership. Because the contract is based on a firm fixed price, there is a cap on what the government can pay the contractor, and costs beyond that level are paid 100 percent by the contractor. On your side, however, there is risk that even after you devote time and effort to working with the contractor, the project may fail.

QBS: Qualifications Based Selection

This is a three-stage process as described below.

1. You begin by preparing a general statement of work (which is far less specific than the one discussed in the RFP process), a schedule for selection, and evaluation criteria. Then, a request for statements of qualifications (SOQ) is published. The responses are received and evaluated by a selection committee, which interviews a small number of candidates and ranks them.
2. At this point, you and the top ranked consultant begin to define in detail the scope of services (which is better than creating the scope on your own). If necessary, a pilot project can be implemented to further crystallize what the project will entail. A fee proposal can then be prepared.
3. You can either accept the proposal or refine the project scope to bring the cost and schedule more in line with expectations. If an agreement is not reached, you have the option of partnering with the consultant ranked second.

None of these options is a magic bullet, however. You may find that one will work wonderfully for your organization as is, or you may have to make modifications to improve results. Then again, there is nothing to stop you from creating an entirely new procurement approach (except perhaps your attorney).

Key Risk Areas

Change Control

Also called "scope creep," this phenomenon refers to the temptation to add attributes, data layers, or support one more application as a project progresses. Collectively, additions can doom a project, because they can delay milestones and swallow precious resources. You must proactively manage the data conversion project to prevent scope creep. If an additional attribute or layer is crucial to the success of the project, you have several options.

- Remove less critical attributes or layers.
- Allocate additional time, money, and staff to accommodate the request.
- Draw a "box" around the current project and file the request under unfunded requirements.

An effective method for avoiding scope creep is to prepare a thoughtful, documented plan for the conversion project. Give participants an opportunity to add requirements at the beginning, and then pare the project down to essential layers and attributes based on funding, schedule, and staff requirements. Additional requirements should only be included if you determine they will benefit the project.

Guarding Against "Foolproof" Claims

Most GIS software packages have the capability to import a variety of published data formats, especially those formats from federal agencies (e.g., digital elevation models and digital line graphs). When these formats are used, conversions are supposed to be foolproof. Some software products also allow users to work with and analyze data in their native data format without saving the data in the software's own format. This feature allows CAD and GIS-formatted data to be used in spatial analysis without worrying about saving outdated data. You should avoid saving data layers in a variety of formats, because it can confuse users about which layers are correct. For more information regarding transfer standards, see Chapter 15, "Spatial Data Transfer Standards."

Database Design Parameters

Once you determine the geospatial framework for your data, all collection efforts must use the same datum, ellipsoid, and map projection whenever possible. If data must be collected in another framework, the data must be converted for use with the database design.

For example, one agency wanted to collect GPS data of nesting sites for a particular endangered species to plan new logging roads for a forest harvest project. The agency did not specify the datum to be used, and it failed to verify the data collection used by contractors relied upon the same geospatial framework used by the agency. When it came time to apply the data, the agency mapped roads with a 200m buffer around nesting sites. However, the data were inaccurate because of the different geospatial reference systems, and the agency ended up putting a logging road 15m from a nesting site. Luckily, no trees hosting endangered species were removed.

When using data from other sources, verify that the information is in the same geospatial reference system you are using. Working with data concurrently that have different projections or datums leads to questionable spatial analysis results. The data must be converted before they can be used simultaneously.

Data Quality

Also known as data confidence, the quality of your data is determined by the positional and attribute accuracy of the data set. If you acquire data from a federal agency with a map product mission, the data will comply with the National Map Accuracy Standard. (For more information on the standard, see the USGS Internet site at *http://mapping.usgs.gov/www/html/1stand.html.*)

The confidence you place in your data translates into confidence in the results of your spatial analyses. Errors and inaccuracies tend to multiply during modeling efforts, compounding data inaccuracies. No one has perfect data; they do not exist. However, as data collection techniques improve, data quality is improving—if you are willing to pay for it. Another important milestone in data quality is the adoption of the Federal Geographic Data Committee's Content Standards for Digital Geospatial Metadata. This effort encourages professionals (and requires federal agencies) to document their new geospatial data. For more information on data documentation, see Chapter 5, "Documentation of Data Conversion." Data quality control and quality assurance are addressed further in Chapter 16, "Quality Control/ Quality Assurance."

Cost and Schedule

Cost and schedule are two of the most difficult and risky components in a GIS data conversion project, particularly in the planning stages. After weighing their options, many organizations find it cost effective to hire data conversion firms to help allocate funds and set a schedule. In many cases, including vendors in the planning stage can reduce risk.

Project cost and schedule, of course, change for each agency and each project within an agency. The availability of many project components—hardware, software, data, personnel, and other resources—can shift. It is possible you will draft project cost and schedule many times before all staff agree. Keep in mind that all figures and dates should be considered subject to change until the pilot is complete. At that time, you may want to revise costs and schedule to ensure the project's success.

The project's time frame will reflect your organization's priorities as well. Some organizations, preferring quick results, will fund a project fully at the outset. Others are nervous about committing substantial resources to a new project and may opt to proceed slowly in order to gain experience.

Communication

As with many endeavors, communication is key to a successful conversion effort. This does not mean obtaining hourly updates, but determining the appropriate interval between update reports. Determine what works best for your team. If a major problem occurs, alert your supervisor immediately. You can minimize damage if you respond to problems promptly. Also remember communication sources outside your team. Others who may have dealt with a similar problem, available through e-mail lists or in chat rooms, may have a workable solution for you. These individuals may even have programming scripts or tricks to help you on your way. However, be cautious with foolproof claims, particularly when using code from outside sources. Your datum or network configuration may not be compatible with workarounds.

User Acceptance

User acceptance translates into user confidence in the data. Including end users in the process ensures they will accept and have confidence in the final product. End users know the types of checks and balances they want performed on their data. Checks may include data verification via GPS location checking, check plotting, attribute relationship checks, visual surveys, or vendor tools. The project plan should contain the criteria and checks for data validation.

Many end users know which datum and map projection will work best for respective projects. Together with database design experts, end users can determine an optimal database design. A team approach ensures that you draw on more expertise when planning your project, and increases the chances the pilot will succeed.

Data users are valuable sources because of their familiarity with the geospatial extent of the conversion project. This characteristic means they are excellent quality control/quality assurance evaluators. They are well acquainted with the "lay of the land" and can identify errors in data location and their attributes, often simply by viewing the data on a computer screen.

Data Maintenance

Data maintenance is a high risk area because most organizations are behind on hardcopy updates of their data even before they begin a data conversion project. Adding digital data to your maintenance is often more than one agency resources can handle. To boost the importance of data maintenance, keep this in mind: your data are only as accurate or up-to-date as you make them. Once you miss the first scheduled update, your become data old and your analyses may be incorrect. For example, lives may be at stake if the data are used for 911 activities. For more information on data maintenance, see Chapter 16, "Quality Control/Quality Assurance."

Management Support

Management support often determines the success or failure of a project. Anyone who has read Dilbert cartoons understands this reality. Managers are necessary to project success because they play a vital role in planning, organizing, coordinating, directing, and controlling the work environment. Management must be kept informed of the good news—and the bad—related to your project. Always keep your managers informed.

✓ **TIP:** *Always have good news to offer those managers who inquire about the project, but also be sure to deal firmly with problems.*

A good relationship between you and your manager ultimately depends on you. You have much to gain from a good working relationship; managers can provide guidance and eliminate difficulties. Managers can allocate funds and personnel if your project is thrown off schedule by events beyond your control. If you trust the managers involved in your project, their power and experience can help make your conversion project a success. On the other hand, a poor working relationship with your manager can only hurt you. The support you might receive could count doubly hard against you if you neglect the relationship and a manager becomes averse to the project.

Procedural Innovation

GIS and computer experts who automate functions can improve the work environment and make you more effective. This makes new methods and innovations extremely attractive. However, use them wisely and with caution in a data conversion project. Always recognize the usefulness and constraints of a given method. For example, if a method was developed to convert line data, it may not be valid on polygon data sets. Make sure that you check the method for errors before using it on the entire data set. Likewise, if you develop a method you feel is sound, let others know. Write a white paper on the experience, send e-mail to list servers of appropriate topics, or present a paper at a conference. The skills you gain will foster your professional development.

Mitigating Risk and Liability

Liability and risks related to GIS data have encouraged many data providers to include liability statements in the data they distribute. To counteract potential liability issues in your conversion project, address them as you plan the project. First, review liability actions against similar organizations to determine what went wrong. By examining others' bad luck, you can strengthen your plan and protect yourself from litigation.

One poorly designed data development effort became a deadly mistake when a company supplied flawed flight approaches for airports in mountainous areas. Because of the flaws in the data, fatal accidents occurred. This liability case is currently working its way through the court system.

Another step you can take to protect your organization is to involve an attorney in the planning process. This is an expensive step, but it could save you millions of dollars in liability suits down the road. Because data liability is a relatively new field, however, verify that the attorney has a background in the subject or in class action litigation at a minimum.

To be sure, the best way to mitigate risks is to have a sound conversion plan. Include in the plan sound quality assurance/quality control procedures. Document the methods you plan to use and errors you encounter. Errors may include missing attributes, node errors, label errors, and data shifting. Use the pilot project as it is intended, not as the first part of the project. Document the scale and accuracy estimates of the data using a data documentation schema such as the FGDC Content Standards for Geospatial Metadata. Ultimately, all you can do is follow your plan and honestly document data integrity.

Conclusion

The key to successful project management of a data conversion project is careful and intelligent planning. The planning stage will help you identify issues relevant to your project and help you anticipate potential trade-offs related to resource or logistical constraints. Ultimately, a solid data conversion plan increases the chances that your project will succeed.

Acknowledgments

OnWord Press thanks the National Academy of Public Administration Foundation for granting permission to reprint portions of the article "Information Technology Procurement: Moving from Contracts to Partnerships," appearing on the Web site *http:/www.clearlake.ibm.com/Alliance/clusters/crt/proadmaj.nab.html.*

Documenting a Data Conversion Effort

Marty McLeod

When you undertake a GIS data conversion project, you must address many questions long before you begin converting data. As part of this process, mountains of documentation are likely to pile up, such as needs assessment surveys; conceptual, physical, and application designs; and strategic, conversion, implementation, and quality control plans. The list of components to outline and document seems endless. This chapter provides insight into how to create and organize documentation to facilitate conversion.

If you have been awarded the responsibility of ensuring the integrity of converted data, you likely appreciate the value of thorough documentation. However, you may also agree that enduring a root canal is more pleasant. After all, writing and maintaining the "what," "why," and "how" for a data conversion project can be time-consuming, frustrating, boring, and tedious.

Benefits of Documentation

The greatest challenge of a GIS project is ensuring that you have reliable data in your database. None of the benefits you hope to realize from your GIS will materialize unless the data you convert are complete, accurate, and appropriate for end users. As such, always remember that—regardless of what anyone says or promises—obtaining reliable data takes a substantial amount of time. The data conversion phase of a GIS project can be long and arduous, depending upon the quality of your sources, the complexity of your database design, and the level of positional accuracy required. Whether your organization uses a consultant or existing staff, having comprehensive documentation that explains your database design, conversion procedures, and quality assurance steps will help to ensure a successful project. Certainly, consultants and vendors appreciate having documentation and they may even assist in preparing it. In particular, having a document that both parties participate in—and agree to—can help resolve problems that may arise during conversion.

Benefits pertain to in-house conversions as well. Unforeseen problems such as losing important team members may lead to lengthy and costly delays. Reliable documentation can reduce the learning curve for new staff, offset some effects of staff turnover, and help substantiate the actions you take that may attract undue attention down the road.

Trying to ensure that every feature you import into your GIS is 100 percent accurate is a sleep-losing, weight-gaining improbability. Data conversion, especially with a vendor, usually adheres to a tight delivery schedule. This reality, coupled with management pressure to prove the investment worthwhile and limited resources to perform quality control, allows little room for error or change. But, problems and changes will occur. For example, if your organization currently uses manual or hardcopy maps, then converting to a GIS can be quite disruptive for staff. People who rely on maps to carry out their daily responsibilities, who have used the same maps for many years, may panic when you try to introduce change. Documenting how and why you made certain decisions during the conversion can help calm those fears when the time comes to integrate your GIS into the normal work flow.

Creating Data Conversion Documentation

Try to keep four things in mind when creating and organizing documentation: simplicity, portability, motivation, and detail.

Keep It Simple

One of the most fascinating characteristics about your designs, source documents, and data is that they will change. When changes occur, the documentation must be updated appropriately. The easier it is to maintain documentation, the more likely you will be to keep it current. One method of achieving simplicity, especially in the database design document, is to avoid referring to specific items in more than one place. Therefore, when change occurs, you only have to update one section. This will also reduce the possibility that your documentation will become inconsistent, which could confuse consultants, vendors, and users.

Consolidation is another method of keeping your documentation simple and easy to use. While GIS data conversion can be difficult, the documentation that accompanies it should not be complex or hard to read.

Make It Portable

Most documentation created for the data conversion phase itself (i.e., source document descriptions and conversion procedures) will be specific to the conversion and will not be needed once the conversion is complete. However, your database design document is far different—it is the one document you should consider a living organism. Its life span is effectively indefinite, because it will be useful long after the conversion is complete. Along the way, it may change in subtle or pronounced ways.

In addition, data conversion documentation is typically created in a linear fashion; one document follows the next (e.g., normally physical database design follows conceptual database design). With these things in mind, you can save yourself valuable time if you aim to create and organize documentation that can evolve instead of the kind requiring extensive revisions. If you rely on a consultant or your GIS software vendor to create your documentation on the database design (or any other aspect of the conversion), make sure it is delivered in a compatible digital format, not just hard copy—particularly if graphics are included. Also, if your organization makes use of intranet technology, you may want to create on-line documentation of your database design in order to distribute updates efficiently.

Address Motivation

Once decisions regarding database content, application needs, and cartographic design have been made, it is wise to document the reasoning behind them. Documentation that explains why certain decisions were made can be useful when your own memory fades. Unless you use GIS technology to replicate exactly what exists in your paper world (which is highly unlikely), you may encounter resistance and questions regarding the choices you made. The value of documentation increases over time.

With proper planning, you will no doubt identify areas where improvements to your map products can be made. Take the time to document design aspects that significantly change how your maps look and perform—or how people perform map related tasks. By doing so, end users will be better able to interpret, accept, and apply the new map products in their work. While this may take time and effort, it can return substantial dividends when the GIS data are integrated into the work environment.

Provide as Much Detail as Possible, and Then Some

While no one sets out to make decisions based on assumptions, it happens frequently and often unconsciously. Taking the time to create detailed specifications that capture your visions and map products can give everyone involved a sound mental picture of what is expected. Because pictures speak volumes, detailed graphic examples of how you envision map features to appear on your maps may be more effective than written descriptions. If you plan to merely replicate existing paper maps, then graphics are probably not necessary, especially if your existing maps become the primary source documents. However, if your GIS design differs significantly, graphic examples can be essential.

What to Include

Six major components that should be addressed in the data conversion documentation or manual follow.

- Project implementation objectives and priorities
- Database design
- Data conversion plan and procedures
- Quality control plan and procedures
- Post-conversion updates
- New stuff "outside the box"

Project Implementation Objectives and Priorities

Creating documentation that outlines the objectives, priorities, and scope of your project can help the project stay on track and catalog new ideas that surface once data conversion has begun. If the comprehensive needs assessment and strategic planning phases were complete and well-documented before you arrived at the data conversion stage, then much of the content for this section can be gleaned from those documents. There undoubtedly will be features and applications that people will want to add during the data conversion stage and, depending on your management support, these new ideas may change project objectives and create added stress. (See the section on change control in Chapter 4, "Project Planning and Management.") This section of your documentation should cover the reasons you chose to convert your data to a GIS, the initial objectives you expect to achieve, and future objectives. Such information can be referred to time and again when issues arise over priorities and objectives, or simply to refresh your understanding of project goals.

How to handle items that distract you from your conversion objective.

Database Design

The individuals performing your data conversion must understand three things in order to provide a high quality conversion: your GIS database design, current source documents, and preferred conversion methodology. Your documentation on database design should convey each item concisely and completely. The examples and suggestions offered here are intended to provide some guidelines for content and organization of your database design documentation. For example, a table of contents might include design concepts and applications, the data model, table outlines, feature descriptions and source examples, and updates/changes.

Design Concepts and Applications

This section describes your conceptual design for data and applications, including an overview of data requirements and special relationships that must be highlighted. This is the place to document the reasons underlying the design, especially when you plan to deviate significantly from the current system.

Data Model

This section addresses the physical design of your data and how it relates to the specific GIS software and/or database management systems being used. Describe in detail how the conceptual design is being realized within the software design.

Table Outlines

This section includes table definitions or other structures derived from your physical design.

Sample table outline for a water valve attribute table.

RDBMS TABLE NAME **valve_type.atr**
PURPOSE stores attributes of water valves related to features in the GIS

Field	Data Type	Field Size	Required	Purpose
tag	number	8	Yes	primary key, ties to GIS feature
installed	date	na	No	date valve was installed
type	number	1	Yes	type of valve (1,2, or 3)
manufacturer	chacter	10	No	name of valve manufacturer

Feature Descriptions and Source Examples

The section on feature descriptions and source examples should describe, in as much detail as possible, each feature being converted into the GIS. Items to document and provide for each feature follow.

- Feature name
- Feature type table name(s)
- Feature description and placement rules

- Attribute descriptions, rules, dependencies, and legal values
- Data conversion source(s) and method(s)
- Source description and example attachment
- Graphic example and attachment

Updates/Changes

This section should record changes, as they occur, with regard to the first four sections. The table below illustrates this process using the bulleted items under the previous section as an example.

Feature name	*Valve*
Feature type	Point
Table name(s)	WATER_VALVES
Feature description and placement rules	Valves are categorized by their use in the water system. Connectivity to water mains and symbology of valves is based on the USE attribute. All valves with USE = MAIN LINE must be located at the node of two water mains. SERVICE valves are located ...
Attribute descriptions	
USE	This attribute pertains to the specific use for the valve within the water system and is specified on the Valve Card and/or Valve Map. The legal values for this item are: MAIN_LINE, SERVICE, HYDRANT, FACILITY.
SIZE	The size in inches of the valve. If source does not indicate size populate with -9. Legal values are: .75, 1, 1.5, ...
Data conversion source(s) and method(s)	Valve Maps, Valve Cards Location of features will be derived from the Valve Map by digitizing. Attributes are keyed in from the associated Valve Card. The number next to the valve on the valve map is the key to the associated valve card.

Source description	Valves are shown on the valve map by a filled circle 1/4" in diameter. The text next to each valve on the valve map is the NUMBER attribute and is the connection to the Valve Card. All attributes to be populated are taken from the Valve Card. See attached source examples for attribute listing.
Graphic example	See attachment.

Data Conversion Plan and Procedures

Your data conversion plan document covers the people, tasks, and procedures involved in data conversion, as well as problem resolution. This outline assumes the use of a conversion vendor because such situations tend to be more complex. Elements of this document would include the production team; source document preparation and tracking; project deliverables, schedule, and terms of payment; and acceptance criteria and error correction.

Production Team

Names, responsibilities, and contact information for each person involved are compiled here.

Source Document Preparation and Tracking

Outline the steps and responsibilities involved with source preparation and ongoing source maintenance. This includes procedures and responsibilities for scrubbing and reproducing all source documents, as well as procedures for tracking updates to the manual system during conversion. Outline procedures and a timetable for resolving source document problems and discrepancies.

Project Deliverables, Schedule, and Terms of Payment

This section formalizes the amount of data, data format, a delivery schedule, and timetable for accepting or rejecting the data. Include a schedule for redeliveries as well. Note payment terms and penalties if data are unacceptable.

Acceptance Criteria and Error Correction

Detailed documentation that outlines the difference between acceptable and unacceptable data can play a vital role in your conversion, particularly when a consultant is involved. Both parties should have a clear understanding of what is expected and penalties for failing to comply. A consultant typically must meet an error threshold agreed to in the initial contract. To avoid misunderstandings, describe in detail what constitutes graphic element and attribute errors, and how errors are counted. For example, a graphic element error could be an annotation more than 10' away from its associated water main graphic, while an attribute error could be a valve coded with the incorrect installation date. Next, describe responsibilities for error correction.

Quality Control Plan and Procedures

Your quality control (QC) plan is designed to document how data accuracy will be verified. The components of a QC plan document are listed below.

- Loading a delivery, and generating check plots and error reports
- Identifying the tools used for quality control
- Documenting and calculating errors
- Establishing features to be checked, and the methods for checking them
- Checking redelivered data

Loading a Delivery and Generating Check Plots and Error Reports

Document the steps to import a delivery into the system and generate check plots and reports used in the QC process. If you use automated software routines for data verification, outline the procedures for running the software.

Identifying Tools for Quality Control

Outline the tools (i.e., check plots, reports, and software) used in the QC process and their purpose. If you create check plots or report dumps, describe the features displayed in them.

Recording and Calculating Errors

Describe procedures for documenting and calculating errors.

Checking Features

A list of all features being converted and how to perform QC should be addressed in this section. Include the tools, sources, and steps used to ensure accuracy. The table below illustrates this step.

Feature	*Valve*
Tools used	Valve check plot, Valve attribute report, QC software report
Sources used	Valve Map, Description Page, Valve Card
Items checked and how (location and number, do they reflect source)	Verifying location is accomplished by comparing the check plot against the Valve Map. The number next to the valve on the Valve Map should match the number on the check plot. If valve is not located correctly indicate proper location on check plot. If number is incorrect cross out the number and write correct number on the check plot.
Connectivity, do they snap to water mains properly	The QC software will check that valves connect properly to water mains. If the QC software report identifies valves that are incorrectly connected then circle the valve on the check plot and write "Connectivity Error" next to the valve.
Attribute values, do they reflect source	Attributes are checked by comparing the Valve Card with the Valve Attribute Report. Document errors by writing the correct value next to the incorrect item on the Valve Attribute Report.

Checking Redelivered Data

If a data delivery is rejected and subsequently redelivered, then it must pass through your QC process again. Depending on the reasons the delivery was rejected it may not be necessary for all data to be rechecked. This section should cover the procedures used to check redeliveries.

Post-conversion Updates

It is very likely that your conversion vendor will be unable to populate all attributes for the features being converted, either because of incomplete source availability or financial limitations. You may also have additional feature types included in your database design that will not be converted for similar reasons. This section should document these cases and other items requiring attention after the initial conversion.

One of the interesting (or frustrating) things about the data conversion process is that some features will not conform to your design specifications. You spend hours (or days or weeks) outlining specifications only to find that subtle nuances in your current mapping standards cannot be duplicated within your GIS design. For example, a water utility conducting a data conversion had a specific design rule regarding a section of pipe, known as a lateral, that feeds water to a fire hydrant. The rule required that every lateral have a fire hydrant at its terminus. About three months into the conversion, the data vendor encountered a source that showed a lateral without a fire hydrant on it, with the text "FH STUB" next to the lateral. In total, the utility had 30 similar cases (dating back to the 1950s) where laterals were laid for future hydrants that never materialized. To comply with the utility's design rule, the vendor had developed software to ensure that every lateral had a hydrant at its terminus. Because it would have been costly to change the conversion methodology at that point, the utility decided to have the vendor add the hydrants, and the utility would remove them once the data were accepted.

New Stuff "Outside the Box"

If you are responsible for data conversion and guiding the direction of GIS in your organization, a section in your documentation on new items could be beneficial. Once a data conversion effort has entered the full implementation stage, new ideas for applications or data deemed outside the primary scope of the project can be documented and reviewed once the initial conversion is complete. You may want to create a simple form to catalog new requests. Forms could be completed by the individuals making the request and then logged in a notebook for future reference. Or, the form used during the needs assessment phase could be easily adapted. The form should include the items listed below.

- The individual making the request or suggestion
- A description of the application
- Data needed to support the application
- Data source(s)
- Who would be responsible for maintaining the data
- The type of output products required

A Final Thought about Documentation

Historically, most organizations involved with mapping have a poor record of developing comprehensive procedural documentation for maintaining their maps and records. While there may be some documentation for graphic drafting standards collecting dust in a file drawer, you would be hard-pressed to find documentation that specifically outlines the steps and procedures necessary for maintaining and ensuring the high quality of an organization's maps and records. Typically, procedures and techniques are passed along verbally as new staff assume the role of map maintenance. As such, maps and records often evolve with an inconsistent appearance and quality. Because maintaining maps and records can periodically become a very low priority for an organization, significant backlogs can develop. Such mental record keeping can ultimately harm the effectiveness of your GIS.

Converting to a GIS is an expensive investment that will pay tremendous dividends only if your converted data are—and remain—complete and accurate. The quality of analyses and decisions derived from your GIS rely entirely on the quality of your data. To help protect this investment, your organization should develop and document procedures for maintaining and ensuring the quality of data within the GIS.

Of course, even with the most ambitious planning, procedures, and documentation, the toughest roadblock to fully integrating and maintaining a GIS will be personnel. Individuals accustomed to traditional methods may have difficulty accepting change, which can be frustrating for project managers. However, concise documentation may be the first step toward helping such individuals understand the value of GIS.

Acknowledgments

Thanks to Pat Hohl for offering me the opportunity to be involved with his book. His support and encouragement to take on new things over the years has benefited me greatly.

Section 3: Understanding the Target System

Chapter 6 focuses on the components of a GIS target system (hardware, operating system, networking, and GIS related equipment), major players in the GIS software industry, and database design.

Understanding the Target System

Atanas Entchev

Most successful endeavors require a clear understanding of project goals, and GIS data conversion is no different. The data must be converted into a format compatible with your desired GIS, or target system. To understand the process, you must take into consideration the type of system you have or envision, its hardware and network specifications, and operating system; the GIS software you choose; database design; and spatial accuracy. This chapter covers the basic principles underlying each issue.

Different Levels of GIS Use

In its relatively short existence, GIS has evolved significantly in the role it plays and the way it interacts with other organizational activities. The first GIS applications were projects—an agency or department with a need for spatial analysis would implement GIS to solve a specific problem. After the spatial problem was solved, the department ceased its involvement with GIS. Project based GIS continues to dominate many organizations today.

However, GIS is already evolving to the next level, departmental GIS, where it is an ongoing activity and often the backbone of entire departments. For example, the New Jersey Department of Environmental Protection (NJDEP) has a very strong GIS department that plays an increasingly significant role in how the NJDEP carries out its duties.

This leads to the next level, enterprise GIS. In this case, an organization (or enterprise) implements GIS technology on a regular basis and throughout its activities, making GIS an intrinsic part of its operations.

Many GIS observers believe the industry is heading toward societal GIS in which local governments will be able to provide public services on line with GIS and receive feedback. Some prototypes are already in place at this time. Examples include the City of Los Angeles (California) World Wide Web GIS project, and the City of Ontario (California) GIS World Wide Web site.

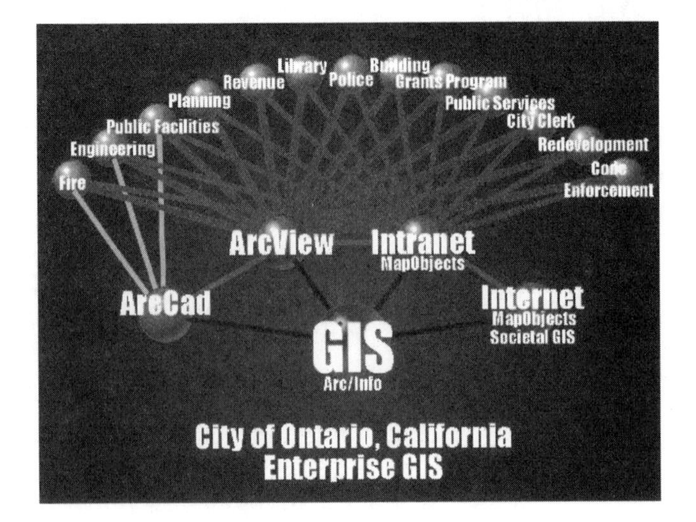

Ontario, California's Web site, an example of societal GIS.

Computer Networks

Because GIS involves computers, all issues relevant to the computer revolution also apply to GIS. Today, more and more computers are interconnected. In what has now become SUN Microsystems' official motto, CEO Scott McNealy once declared, "The network is the computer."

This means you must be knowledgeable about all networks, network protocols (the languages computers use to communicate with each other), and network operating systems. The opposite of the networked environment, the standalone GIS station, is covered later in this chapter.

First, you must consider the different types of networks and network operating systems. In general, networks differ according to their layout, or topology (not to be confused with GIS topology, which is discussed later in this chapter). Network types include peer-to-peer, file server, the much-hyped Internet, and Internet offsprings, the corporate intranets. There are also local area networks (LANs) and wide area networks (WANs), which differ only in the areas they cover, not network topology.

Networks connect computers using a sophisticated web that allows users to share resources such as data files and printers. The network appropriate for you depends largely on your applications and your organization's structure and operations.

Computer networks consist of servers (that share files, e-mail, printers, and other peripherals), workstations, and network connections. A computer network usually has one or more servers hosting data files. Workstations often have applications software installed locally and connect with file servers to retrieve data.

How do network operating systems differ from each other, and how do you determine the right one for you? It is difficult to evaluate target system components separately from the rest of the system. Factors such as computer operating systems, GIS software, computer hardware, and the type and amount of data all play important roles. In addition, some network operating systems double as computer operating systems, while others run the network exclusively.

Network Operating Systems

Each network operating system (NOS) exhibits several basic distinctions. Popular network operating systems include UNIX, Novell, Windows NT, and Lantastic. While the Internet is considered a network in the strict sense, it has a number of characteristics that make it unique because it connects a large and unknown number of computers to each other through a language called TCP/IP (or transmission control protocol/Internet protocol).

Lantastic

Lantastic is a peer-to-peer network operating system suitable for small operations. Such a network does not have dedicated servers; instead, all computers are "peers." With the introduction of the Windows 95 operating system, a peer-to-peer networking capability comes bundled with the operating system. This, and the general tendency to move away from peer-to-peer networks, may account for the decline in Lantastic installations in recent years.

UNIX

UNIX, a very robust computer operating system and NOS, was developed by AT&T Bell Laboratories (before they split to form Lucent Technologies) primarily for scientific research purposes. Its origins led to the two most important characteristics of UNIX: its robust nature and its extreme lack of user-friendly features. UNIX systems are very powerful but not very easy to administer. Traditionally command driven, in recent years several "point-and-click" add-on environments have been developed for UNIX. Such environments make UNIX friendlier and easier to manage.

Novell

Probably the most versatile NOS, Novell enjoys the greatest hardware and software compatibility. Despite its superb technology and large installed base, Novell has seen a decline recently in market share. The reason? The advent of Windows NT.

Windows NT

Windows NT is emerging rapidly as the NOS of choice for small and large agencies. This is not merely a function of Microsoft's heavy marketing. Windows NT packs plenty of nice features, stability, robustness, and excellent security. Add Microsoft's years of experience in the development of user interfaces, and you have a user-friendly NOS unsurpassed by any other system.

The Internet

The Internet connects computers and networks throughout the globe via fiber optics, satellite signals, telephone lines, and radio waves. With an estimated 20 million computers connected, and the number growing exponentially, it is easy to see why Internet talk has become ubiquitous and is not confined to computer geeks. The Internet has finally become mainstream.

The computer industry—which launched its own paradigm shift—is experiencing another because of the Internet. There is talk that computer networks will evolve to become a major medium, perhaps *the* next medium. At present, the Internet is popular because it provides unprecedented ease and accessibility in distributing data and applications—and GIS is all about data, information, and knowledge. It is essential that every GIS professional and manager be familiar with the way the Internet works, because of the strong impact this new medium is already having on the field.

Today, GIS data sources can be retrieved across the Internet from multiple locations to perform a single GIS overlay on the desktop. Or, you can access map server sites on the World Wide Web to create simple maps without owning or installing GIS software locally.

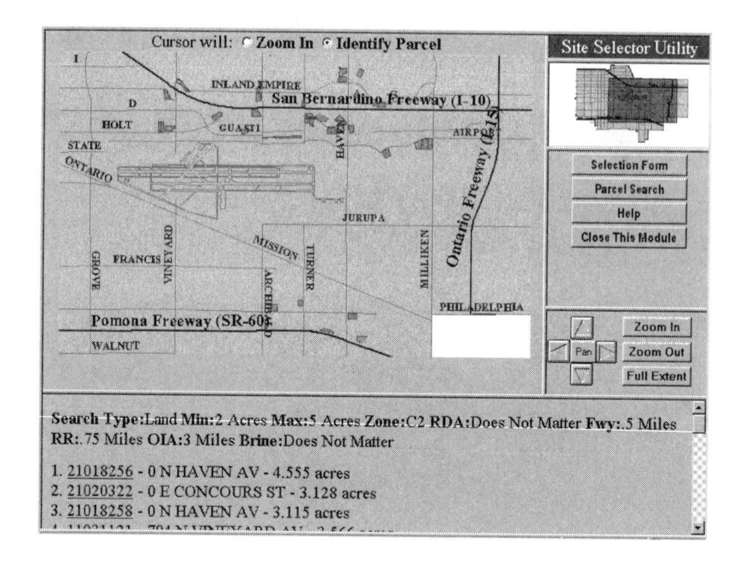

Ontario, California's GIS Web site delivers data.

Network Security

A paramount issue for all computer networks is security. System administrators must provide a secure working environment for network users and be able to restrict access to sensitive data resources. These issues apply equally to the Internet.

The Internet has seen more than its share of security breaches. One reason is that when the Internet was designed in the late 1960s, it was designed exclusively for military and academic communications applications. There were no expectations of malicious usage and as a result, security received attention. All that has changed, however, now that the Internet can be accessed from numerous online services and countless local Internet service providers. Every university in North America—and probably the world—has Internet access. This provides access to innocent hackers and pranksters, and more importantly, malicious users who attack Internet systems and data.

The issue of security has been a high priority since the Internet became a popular public phenomenon. Tremendous effort has been focused on improving security, and today the Internet is significantly safer. Nevertheless, security continues to be an issue and must be taken seriously, particularly because GIS data constitute a substantial investment.

Computer Operating Systems

The computer operating system (OS) is a piece of software that governs interactions among physical computer components and user applications. The OS is like an interpreter coordinating discussions between the computer processor and commands issued by your GIS software. The computer OS is crucial to computer performance. In addition, this topic overlaps somewhat with NOS. Some computer operating systems, such as UNIX and Windows NT, are both computer OS and NOS, while others (e.g., Novell and DOS) are one or the other.

UNIX

The NOS characteristics of UNIX, mentioned earlier, apply here as well. While UNIX is a very robust, very secure 32-bit (and even 64-bit) OS, it remains problematic. Not only does the UNIX OS suffer from an absence of user-friendly features, it is also plagued by different, incompatible implementations distributed through computer manufacturers such as Hewlett-Packard, Sun, DEC, and others.

Disk Operating System (DOS)

DOS was written by Microsoft Corporation's founder Bill Gates for IBM's first personal computer. Gates retained the copyright to DOS, and when he and IBM parted ways, Microsoft launched its version of DOS, or MS-DOS. DOS has evolved from its first release (1.0) to its most recent version (6.22). (DOS version 7.0 for Windows 95 is covered later.)

For many years, DOS was a very popular and reliable OS for the PC. Many GIS applications were initially written for MS-DOS. However, the demise of DOS is virtually guaranteed because of its lack of multitasking capabilities. DOS allows you to run one and only one application at a time, regardless of how fast or powerful the computer.

Windows

The need for multitasking was addressed by Microsoft when it introduced Windows. Windows 1.0 through Windows 3.11 for Workgroups were not "real" operating systems, but rather operating environments—add-ons that required MS-DOS in order to function. Windows provided multitasking capabilities on the desktop along with a graphical user interface (GUI), or a point-and-click interface.

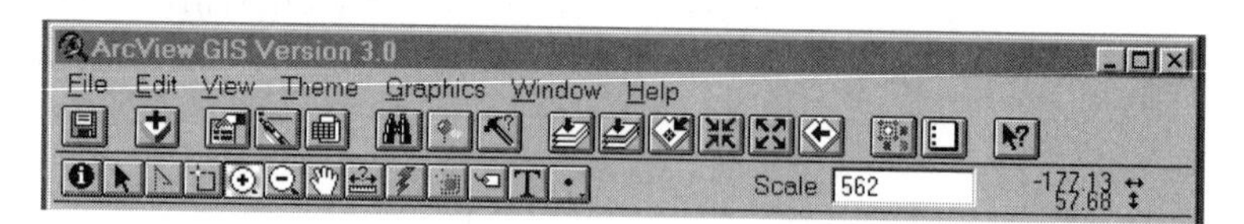

A GUI in ArcView GIS.

The first releases of Windows were notorious for their instability. They would crash frequently for no apparent reason. Microsoft has come a long way since then, and the latest version of the Windows operating environment, Windows for Workgroups 3.11, boasts significant stability. Windows has also become the platform of choice for most software developers.

As of version 3.11, Windows was still missing a few features, however. Because it was not yet a true operating system, and DOS was a 16-bit operating system, Windows had to be as well. Today, an increasing number of software developers are writing 32-bit applications, because of the increased speed and other features such architecture affords.

If Microsoft wanted to compete with the 32-bit UNIX, it had to provide its own 32-bit operating system. Microsoft's answer was to provide two: Windows 95 and Windows NT.

Windows 95

Windows 95, a 32-bit operating system, does not require DOS. However, because of the huge installed base of DOS-Windows 3.x applications, Windows 95 was designed with backward compatibility in mind. Thus, it includes some 16-

bit DOS code in it, rendering it a "non-pure 32-bit OS," according to software purists.

This code, however, is quite beneficial. If Windows 95 were truly 32-bit, there would be no smooth transition from DOS-Windows 3.x to Windows 95. Your old applications would not run under the new OS. Because of Windows 95's impurity, it runs almost all DOS applications, and virtually all Windows 3.x applications, while adding stability previously unknown to Windows 3.x users.

Windows NT

Windows NT is Microsoft's answer to UNIX, and addresses most, if not all, of the shortcomings of DOS-Windows 3.x and Windows 95. Some of its most important features: airtight security, and a truly 32-bit architecture.

Such functionality comes at a price, however, and it is not merely financial. Because of its true 32-bit architecture, Windows NT lacks the backward compatibility of Windows 95. Thus far, it also lacks broad support from hardware and software vendors.

All of this is changing. In an industry where major breakthroughs are announced every few weeks, it will not be long before Windows NT enjoys the popularity and broad vendor support typical of Windows 95. This is especially important in GIS, where major software vendors have started to port applications from UNIX to Windows NT.

Macintosh OS

The Mac OS, the operating system for Macintosh computers, accounts for about 4 percent of the computer market, according to CNET, Inc. The Mac OS and Apple Computer, Inc., enjoy incredible devotion among their followers. From Mac users you will hear that Windows 95 is where the Mac OS was in 1987. Indeed, the Windows 95 interface seems to be fashioned after that of the Mac. But Apple did not invent the GUI, either. It was first implemented at Xerox Corporation, and Apple quickly embraced the concept.

Apple Computer seems to be in dire straits. Shrinking market share and huge quarterly losses are bad omens for an otherwise excellent technology. From a GIS prospective, more and more software vendors will likely discontinue Mac product lines, or add Mac support months after Win-tel (or Windows-Intel) releases. Still, only the future can tell what will happen to the Mac OS.

Mainframes

Mainframe computers, or Big Irons, were declared dead a few years ago. Today, they seem to be making a comeback. The mainframe computer model was distinctive for its powerful number crunching abilities, with a number of "dumb" terminals connected to it. However, only the model (not the computer) was declared dead. Although many mainframes were replaced by "client servers" (in which terminals became PCs) several years ago, the advent of the Internet and World Wide Web has fostered interest in this model again. The components may have different names, but the concept remains the same. What used to be the mainframe is now known as the "Internet server." Dumb terminals are now called "Net PCs" or NCs.

Web Browsers

What are Web browsers, and why mention them with other operating systems? If you believe Marc Andreessen of Netscape Communications Corporation and author of the first graphical Web browser, Mosaic, the Web browser will be the next operating system. Sound farfetched? Perhaps, but consider what the most advanced Web browsers can do today, and you may have second thoughts.

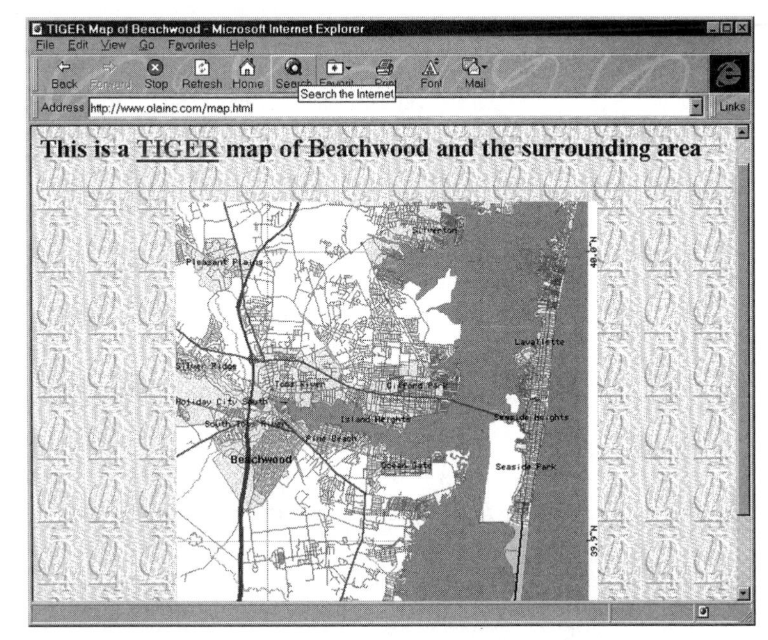

A Web page accessed using the Microsoft Internet Explorer browser.

A Web browser is software that allows users to access the World Wide Web. The most sophisticated browsers today are Netscape Navigator and Microsoft Internet Explorer. Both pack serious features that transform them into far more than tools for merely browsing.

The Web is already home to many interactive mapping applications that can turn your browser into a dumb GIS terminal. But once you can launch an application from within your browser (functionality that already exists in Navigator and Explorer), the browser moves one step closer to an operating system. The conceptual operating system of the future may be very different from today's understanding of what an operating system is and does.

JAVA and ActiveX

The above discussion leads to the products Java and ActiveX which, at their simplest, are programming languages for the Internet. The idea behind them is that programmers can write applications to be read and run by your browser. The novelty here is the concept of "run once, run anywhere," a trademark of the Java development team.

Therefore, proponents say, if you write a Java application, all you need to run it on your computer is a Java compatible browser. The traditional operating system goes to the background. Regardless of whether you are running a Mac, UNIX, or Win-tel machine, the Java application will run for you as long as you have a Java compatible browser. Does this mean the browser will become the OS, or perhaps an extension of it? If you recall the relationship between Windows 3.x and DOS, this notion may sound familiar.

In short, several trends appear to be shaping the development of computer operating systems: GUIs, security, support for 32-bit applications, and multi-tasking. With these issues in mind, you can begin evaluating operating systems. Remember that no component of your GIS should be viewed separately from the rest of the system. While critical to the success of your GIS, the operating system must be in sync with applications software, hardware, existing and future digital data, operating procedures, and users.

Computer Hardware

Computers are composed of one or more central processing units (CPUs); storage, input, and output devices; and networking connectivity devices. When selecting computer hardware, look for speed, reliability, compatibility, and the ability to upgrade.

Processors

The processor (or CPU) is the number crunching component, or heart of your system. A CPU's internal speed is measured in megahertz; the more megahertz, the better off you will be. The speed of your processor and the amount of random access memory (RAM) determine the overall speed of your system.

While specific recommendations are beyond the scope of this chapter, greater CPU speed is always better. When you acquire a new computer system, always buy the fastest processor you can afford rather than budget for an upgrade later. Some upgrades are simply not worth it because you will not see a leap in performance comparable to the money you invest. Moreover, while RAM can be easily added in properly designed systems, it may not improve your system's performance either. Find a balance that takes into account your needs and software applications.

Computer Chip Manufacturers

Several significant players dominate the computer chip market: Intel, Cyrix, K6, SPARC, Mac/Motorola, and Alpha/ Digital Equipment Corporation (DEC). Intel by far leads the market in computers built for DOS and Windows OS. Cyrix and K6 are second to Intel. If you are considering Cyrix or K6, make sure your GIS applications have been tested and will run on these processors.

SPARC is a processor designed for the UNIX operating system. As mentioned previously, no component in a GIS can be evaluated alone. In this case, your choice of operating system dictates the selection of your CPU, or at least limits your options. The same holds true for the Mac OS, which is powered by Motorola chips.

✓ **TIP:** *GIS application selection drives your operating system selection, which in turn drives hardware selection.*

Storage Devices

Storage devices store digital data and allow for quick and easy data retrieval. There are several different types of storage devices, generally classified as magnetic and non-magnetic media. Magnetic media represent more traditional storage devices, while non-magnetic media are the result of more recent technological developments.

Hard Drives

The classic magnetic storage device is the hard drive. A magnetic disk encased in an airtight container permanently attached to your PC, the hard drive's capacity is measured in megabytes (Mb) or gigabytes (Gb); one Gb equals 1,000 Mb.

Most computer systems use hard drives as the primary storage device for day-to-day operations, and large capacities are common.

Floppy Disks

The use of floppy disk drives is declining. Floppy disks (i.e., removable magnetic disks with reduced capacity) were used in the early days of computing for data transfer, software installation, and storage. The original 5 1/4" disk drive is defunct. Its successor, the 3 1/2" floppy, will soon follow, outpaced by CD-ROMs, portable hard drives, and network file access. At present, more reliable devices with far higher capacity, such as magnetic tapes or optical drives, are used for backups.

Magnetic Tapes

Magnetic tapes have been the medium of choice for backup and archiving for years. Tapes come in many different sizes and formats, and can hold 250 Mb to 8 Gb of data. They are very convenient for unattended (overnight) backups. However, data transfer to and from tapes is not very fast, which is why they are used mostly for archival purposes.

CD-ROMs

With the increasing size of professional software applications, distributing them on floppy disks became inconvenient, so software developers started to distribute products on CD-ROMs, or compact disc-read only memory. CD-ROMs are also an inexpensive and convenient way to distribute large amounts of data.

Another emerging breed of storage device, so-called "portable" hard drives, can hold surprisingly large amounts of data. They are small, portable, and convenient. It would not be surprising if either the portable hard drive or CD-ROM replaces the $3^1/2$" floppy disk drive as the standard.

Input Devices

Every computer system requires input and output devices. Input devices are used to input data into the computer and issue commands to the operating system and the software applications. Output devices are the means through which the computer talks to the user.

Keyboard

The classic input device is the keyboard. A direct descendant of the typewriter, computer keyboards still follow the original QWERTY layout.

Different manufacturers are experimenting with different keyboard layouts, and some of them are very clever. However, for straight ASCII/alphanumeric input, the classic keyboard is likely to remain the standard for the foreseeable future.

Mouse

Graphical user interfaces, or GUIs, have become the interface of choice for software developers and users alike. Instead of typing commands, users "click" on GUI buttons using a "mouse." The mouse, the most popular pointing device, is moved around on the "mouse pad" by the user, and the cursor indicates the mouse's position on the screen.

Digitizing

Input techniques for GIS are covered in detail in Section 5 of this book, "Data Conversion/Input Methodologies." In the GIS field, mice are not just pointing devices. Together with digitizers, they often are used for graphical input into the

GIS, or digitizing. "Heads-up" digitizing is a method in which the user traces an image such as an aerial photograph on the monitor, and adds lines into the GIS by tracing features on the photograph and clicking the mouse. Another method for digitizing is "heads down," where a hardcopy map or paper manuscript is mounted on a digitizing tablet. The user then traces the map features with a digitizing puck, and clicks on buttons to enter data into the GIS.

Electronic Pen

Another fairly new input device is the electronic pen. Used in lieu of a mouse, electronic pens are the best companions for optical character recognition programs, which convert hand-written text into ASCII code. This technology seems to be a particularly promising development for field data entry.

Global Positioning Systems

Another way to enter field data directly into your GIS is using a global positioning system (GPS). GPS receivers receive signals from U.S. Department of Defense satellites. Those signals, containing information about the receiver's horizontal and vertical location, can be input directly into a GIS. See Chapter 12, "Global Positioning Systems," for more information.

Scanners

Graphical data can also be entered into a GIS with scanners that optically convert hardcopy documents into raster digital images. The usefulness of scanning documents into a GIS is the subject of some industry debate. Refer to Chapter 13, "Scanning," for a discussion of situations where the technology is appropriate.

Output Devices

Monitor

Output devices are the computer's way of communicating with the user, a typical example of which is the monitor (it may resemble a television set, but it is far more expensive). For graphical applications such as GIS, the larger the monitor, the better. However, be aware that prices grow exponentially as monitor size increases. Better monitors also have better resolution, measured by the dot pitch. The smaller the number, the higher your monitor's resolution. Today, a good monitor has a .25 dot pitch.

➾ **NOTE:** *Monitor size is measured in inches diagonally across the screen. But watch for manufacturer fine print, which may discuss the monitor's "viewable area." In this scheme, 13" becomes 15", 15" becomes 17", and so forth.*

For high quality graphics rendering and fast redraw times, you will also need a system with a good video card and a substantial amount of RAM.

Printers and Plotters

If you need hardcopy output from your system, you also will need a printer, plotter, or both. Such output devices for generating hardcopy documents vary greatly by size, resolution, and cost.

Printers are usually desktop devices that produce small documents—from 8 1/2" x 11" to 11" x 17". Depending on the type, they are classified as dot matrix (very low quality; virtually obsolete), inkjet (affordable color printing), laser (professional quality black and white or color printing), or thermal wax printers (professional quality color printing).

Plotters are larger devices with standard output up to 36" wide. Early pen plotters, fashioned after a human drafting with a pen, have been largely replaced by inkjet plotters, which are faster, less expensive, and can handle raster data. Electrostatic plotters are high end professional color output devices.

Network Connectivity Devices

Your system connects to a computer network through a network connectivity device, such as a network card, ISDN adapter, cable modem, or regular modem. Each device secures your system's connection to the rest of your workgroup.

GIS Software

The success of your GIS depends on many factors, the most important of which may be the GIS software you select. Software selection can have a substantial impact on your database design and therefore your data. This section covers principal players in the GIS software field, but is not exhaustive.

Non-commercial GIS Software

In the non-commercial GIS software field, GRASS (Geographic Resources Analysis Support System) and IDRISI are the most well-known players. Developed by the U.S. Army Corps of Engineers, GRASS is currently supported by Rutgers University in New Jersey. IDRISI is supported by Clark University in Worcester, Massachusetts. While they are fine products, their impact on the overall GIS market is rather minor.

Commercial GIS Software

In the commercial GIS software field, several companies are major players: Environmental Systems Research Institute, Inc. (Redlands, California); Intergraph Corporation (Huntsville, Alabama); and Autodesk Inc. (Sausalito, California). In addition, Smallworld GIS from Cambridge, has recently gained popularity among utility GIS users.

ESRI

ESRI has a comprehensive line of software products spanning the full range of GIS applications, from high end professional applications to desktop GIS, desktop mapping, and business applications. ESRI's flagship product, ARC/INFO, has become the de facto standard for large-scale GIS implementations, including many among federal and state government departments. Large private sector companies also rely on ARC/INFO software to run everyday operations.

ARC/INFO

ARC/INFO is a professional GIS software package with several extensions, and runs on the UNIX and Windows NT operating systems. ARC/INFO has its own programming language, the ARC macro language (AML). With ARC/INFO's 7.1.2 release, ESRI announced the Open Development Environment for ARC/INFO, which lets users develop custom ARC/INFO applications using industry standard programming environments such as Visual Basic.

ArcView GIS

ArcView GIS is ESRI's desktop GIS solution. It supports a wide range of data sources, including CAD drawings, images, SQL databases, ARC/INFO coverages, and more. Its friendly interface makes ArcView GIS a window into mapping and GIS for less technically proficient users and those who use GIS tools only occasionally.

Other ESRI Products

PC ARC/INFO and ArcCAD are desktop GIS packages from ESRI. While not as rich in functionality as ARC/INFO, they provide sufficient GIS features for small agencies at a lower cost. ArcCAD is specifically designed for existing AutoCAD users who would like to extend their GIS capabilities.

MapObjects, a developer's toolbox full of GIS tools, is designed for developers who would like to add GIS functionality to their applications.

The latest additions to ESRI's GIS product line are the ArcView and MapObjects Internet Map Servers (IMS). Both allow you to publish maps on the Web without reformatting data. In addition, Web browsers can pan and zoom, identify features, and print the maps.

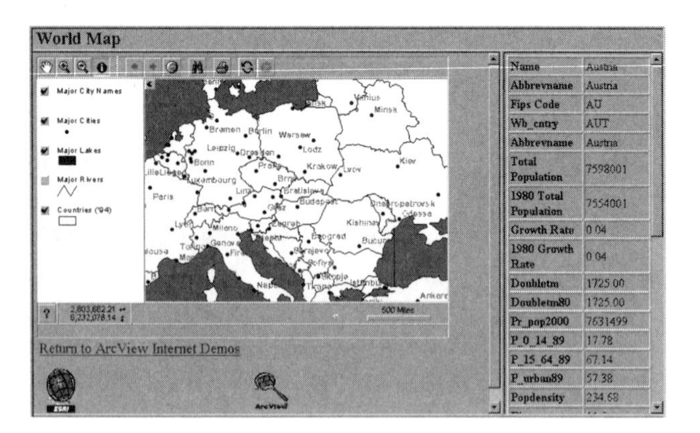

ESRI's ArcView IMS.

Intergraph

Intergraph Corporation is another influential player in the GIS software market. Its flagship product, MGE, and its PC version, MGE PC, are currently incorporated into several "office suites": Civil Office, Municipal Office, Mapping Office, and GIS Office. The suites are designed to satisfy different end user needs. GIS Office, which runs on UNIX, is the higher end product, while Mapping Office, which runs on DOS/Windows platforms, is the desktop version.

Autodesk

AutoCAD Map from Autodesk is the new kid on the block. For a long time, Autodesk tried to present its AutoCAD product, and AutoCAD's ADE extension, as a GIS. The GIS community, however, did not agree that a spatial database management system that lacks topological properties could be considered a GIS. Autodesk's response was Map, an add-on to AutoCAD that handles true topological data structures. Because AutoCAD Map is still new, commentary is problematic.

The most important function in a GIS is the ability to perform spatial analyses. This capability separates GIS from similar technologies such as CAD and desktop mapping. The GIS community believes that the ultimate test of new GIS software packages continues to be whether they handle topological spatial data structures. ESRI's ARC/INFO, PC ARC/INFO, and ArcCAD pass the test. So do Intergraph's MGE and Autodesk's AutoCAD Map.

Other software packages provide "desktop mapping" functionality, such as tools to visualize locations and spatial data distribution, while another set are designed for computer-aided drafting (CAD) and can produce site plans and similar digital drawings. Such packages are certainly useful, but true GIS they are not.

Database Design

The database design has a major impact on your GIS project inasmuch as it determines how you will organize your data. Designs can be optimized for parameters such as storage space, application convenience, access speed, or ease of conversion. There are many trade-offs to take into consideration, not the least of which is the ease of populating the database initially (or converting data into the required database format). You must thoroughly understand the database design to formulate a workable conversion strategy, and you must consider the effect the database design will have on your conversion project. A watershed issue is choosing between raster and vector data. However, often the type of data required for your applications determines whether to choose raster or vector data, in which case your database design is determined for you. In any event, developing a database design is a balancing act. (For more detailed definitions of raster and vector data, see Chapter 2, "Data, the Foundation of GIS.")

Do you need a raster or vector based GIS? The answer depends on the applications you will use. Raster and vector GIS complement each other nicely if you can afford both, and many systems handle both types of data. In addition, conversion between raster and vector formats has become faster and easier, thus eliminating the need to commit to one type exclusively. (Although attribute conversion remains problematic in many cases.)

Digital images, or digital orthophotographs, comprise a special kind of data set. A type of raster data, they typically serve as backdrops for vector data layers.

Topology, mentioned earlier, is a key distinction separating true GIS software from wannabes. Topology is a branch of mathematics that studies spatial relationships. When applied to digital spatial data structures, it adds intelligence to the spatial data set, and allows features to be "aware of" each other. This in turn enables the sophisticated spatial analytical capabilities typical of GIS.

Design Elements

Feature attribution design is very important for a successful GIS implementation. As you know, a basic premise of GIS is the correlation between spatial data (i.e., map features) and attribute data (i.e., textual or other information describing those features). True GIS data formats, such as the coverages in ARC/INFO, maintain a constant and continuing relationship between spatial and non-spatial data. For a successful GIS implementation, some understanding of relational database management principles and formats is useful (see also Chapter 2, "Data, the Foundation of GIS").

There are several widely accepted database formats in GIS applications, including INFO, Oracle, Informix, Sybase, Access, and dBASE. Some of these formats are native to specific GIS applications, meaning a particular GIS can only read from and write to that format. For example, INFO is native to ARC/INFO and ArcView, and dBASE is native to ArcCAD and ArcView. Luckily, GIS users are not limited to only native database formats. The prevalence of the structured query language (SQL) for most databases and among GIS software and data vendors enables GIS users to work with a far wider variety of databases than a single GIS supports.

Design Stages

The conceptual design of a GIS naturally tries to incorporate every player's wish, then squeeze the results into a limited budget. Conceptual designs are often too lofty to be realistic. Yet, they are a critical step in designing your GIS. Letting your fantasies take over is often the only way to foster innovation.

However, the physical design stage must resolve all mundane details and anticipate as many problems as possible. This stage of your GIS design has to elevate reality once again. Compatibility issues must be foreseen and resolved, including compatibility among all components—OS, NOS (if different), computer hardware, GIS software, database software, existing and future data formats, personnel, and established office procedures.

If the physical design stage is given sufficient thought and professional attention, it will pay off in the implementation process. Be prepared for the inevitable surprises, but they will have less impact if the physical design stage is well-conceived. Implementation takes tremendous resolve. Stick to your guns, but also be open to innovations that might improve your system. After all, by now you know that constant change is the name of the game in GIS.

Design Drivers

Why are there so many GIS configurations? Were they designed differently, or did they merely evolve? Both statements are true. Most successful GIS systems are like living organisms and often take a life of their own. But, in the design phase, there are "drivers," or factors influencing your project, that can ensure a long and prosperous life for your GIS.

Design drivers include application requirements, present and future needs, and maintainability. Some of these issues have their origins in the conceptual design stage, while other criteria—specifically maintainability and cost—are deeply rooted in your budget.

How do you design a GIS to be compatible with future developments? Other than staring into a crystal ball, you can ensure that your setup is widely accepted in the industry. This includes OS/NOS, hardware, software, programming languages, and data formats. Do not choose a unique setup just to be different (don't laugh—this has happened).

Next, watch for trends in the computer industry. GIS is a computer technology, and it will always follow closely the general trends in the computer industry. OS/NOS market share shifts are already affecting GIS. Most major software vendors have ported, or are planning to port their applications from UNIX to Windows NT. As the Mac OS loses market share, even dedicated Mac developers such as ESRI might drop their Mac versions. Keep your eye on the developers' languages of choice. Right now, it looks like Visual Basic is a good bet.

Summary

GIS is a very powerful tool. As with any tool, it requires proper understanding to avoid misuse and misinterpretation. Knowing the spatial accuracy of your data is crucial to correctly interpreting your GIS analyses. Likewise, a thorough understanding of the target system is critical to every GIS data conversion project.

Acknowledgments

I would like to thank Pat Hohl for his valuable suggestions and quick help with logistical snags. Thanks to Mike Rottler-Gurley, Christopher Thomas, and Environmental Systems Research Institute, Inc., for providing chapter illustrations. Special thanks to my employer, Owen, Little & Associates, Inc., for ongoing support and encouragement.

Section 4: GIS Data Sources

This section contains information to help you evaluate what types of data might be useful in your conversion effort and where they can be found.

- **Chapter 7** discusses existing sources of data at length and introduces primary sources of new data (e.g., aerial photography, digitizing, COGO, and other methods), and offers guidelines for data selection and preparation.
- **Chapter 8** provides an in-depth discussion of data models and the circumstances under which each model is appropriate. Next, differences between primary and secondary data are considered. Finally, cartographic issues affecting data collection and conversion are outlined.

- **Chapter 9** explains the types of data formats you are likely to encounter in GIS data conversion, provides examples of many types, and suggests sources for obtaining data.

Data Source Types and Preparation

John Kelly

Data Source Types

Implementing a GIS implicitly entails a "database build" phase. The data incorporated into your GIS must come from elsewhere, and are likely to include both internal and external sources. The data may be simple street network layers used for routing. On the other hand, most GIS applications consist of several complex layers of data. In the case of a municipal or utility GIS, these layers collectively comprise what is often referred to as the "landbase." The landbase may include layers of landscape features or layers of parcel boundaries, rights of way, public land, or survey boundaries (also known as *cadastre*, or cadastral data). The landbase is used as a reference to enter utility data features, natural resource features (e.g., soil types, flood plains, or wetlands), or layers of mapping data typical of a city planning department.

Many hardcopy map rooms contain source maps that must be organized before conversion. (MSE)

Whether your database is simple or complex, the data must be compiled and captured from external sources and entered into the GIS. This chapter discusses the sources from which data can be compiled and issues relevant to each type.

Existing Digital Sources

In the infancy of GIS, very little data existed in digital form. As a result, users were forced to capture data from hardcopy sources. Subsequently, however, there has been an exponential increase in the amount and availability of digital data, much of which is marketed specifically for GIS applications. Indeed, when you are planning your GIS database, it makes sense to search for data sources that you can adapt for the applications you envision. This approach is more useful for capturing attribute or tabular data than graphic data because many types of relevant attribute data may already exist in a database such as dBase.

Existing Vector Data Sources

The GIS marketplace abounds with vector data that can be purchased and converted for your GIS. You can either search for such data on your own, or seek the advice of an expert regarding data sources that might be appropriate to include in your GIS. (For more information on vector data, see Chapter 2, "Data, the Foundation of GIS.")

Arguably the most critical factor in determining whether data are appropriate is accuracy. Many data sets may work well in gross spatial maps or applications, such as routing, whose primary focus is not spatial accuracy. For example, several vendors offer street centerline data that essentially consist of street centerline segments, street names, and address ranges. In theory, these data provide a suitable base for routing home deliveries of products or services. However, they are generally inappropriate for use as the backbone for parcel or utility mapping GIS applications.

For several years—an eternity in the field of GIS—the U.S. Geological Survey (USGS) sold maps in digital vector format to the public. Many organizations imported these digital line graph (DLG) data into GIS applications because the data met the spatial accuracy needs for GIS projects under development at the time. For organizations concerned with spatial accuracy at a regional level, the USGS 7.5' quadrangle series saved considerable time and effort. Natural resource agencies made extensive use of these files, as did regional or cross-county transmission utilities. Other sources in this vein are the U.S. Department of Interior's wetland mapping and the Federal Emergency Management Agency's floodplain mapping.

Many large utility companies and some government agencies (particularly at the county level) using GIS have made their data available to other organizations. They market layers of parcel data for use as a landbase, typically on a price per parcel basis. These data can be extremely accurate, a reflection of the stringent accuracy requirements used in building these entities' GIS. Generally, the price of these data is very competitive, and data formats can be imported into your native GIS format and schema with minimal manipulation.

Suitable digital vector data may exist in your own organization. Many organizations possess CAD technology. "As-built" construction drawings, site maps, and fundamental automated mapping files are all candidates for your GIS. However, incorporating such data often requires technical expertise in order to address spatial consistency and suitability of drawings for use in a seamless spatial mapping environment.

Existing Tabular Data Sources

The analytical power of a GIS depends less on the existence of features and more on the richness of attribution related to those features. Regrettably, compiling attribute data can be the most time-consuming and expensive aspect of data capture, because populating these data fields often involves entering attributes for each feature manually into the GIS database. However, you may be able to save time and money if you investigate avenues within your organization to locate relevant digital data files.

Parcel data is one example of existing attribute data. You know that *some* government agency in your region has statutory responsibility for collecting property taxes on parcels and therefore has an extensive database (regardless of whether it is being used with GIS). Once you have obtained these data, the approach for importing them is to key in each parcel's unique identification number into the database, using it as the primary or match key. (See Chapter 2, "Data, the Foundation of GIS" for more information on primary keys.) Then, you can "bulk load" the remainder of parcel data by attaching them to the primary key. A common practice in the industry, bulk loading entails little manipulation of source data, and many GIS software packages can import such data as native functionality. At most, only limited programming is required to extract data from typical ASCII text files, as long as they are delimited in some manner.

Existing Raster Data Sources

Existing raster data sources are often additional components in a GIS. These data are most often associated with document imaging technology, which has matured in parallel with GIS technology. The ability to display and manipulate imagery, an increasingly common support tool for GIS, is now native functionality in major GIS software packages.

Many utilities, engineering firms, public works departments, and other entities have scanned as-built construction drawings into their management systems. Accessing them in a GIS is fairly straightforward using an application that allows you to access, for example, as-built drawings for a particular street segment by clicking on the segment's street centerline. This new dimension of information access is an effective marriage of technologies.

Another example may be found in parcel mapping GIS applications. Many organizations have scanned files of photographs detailing parcel structures and improvements. Many applications now enable you to click on a parcel (or perform another selection) and access the image within the GIS.

Existing Hardcopy Data Sources

The Digitizing Process

Data from hardcopy maps and drawings are captured through the process of tablet digitizing, which can be a very cost-effective method of conversion if your source data are on existing maps. Lower costs stem primarily from lower equipment costs and less stringent requirements for operator training. Linework can be digitized using a CAD station and does not necessarily require GIS software. Digitized linework can be captured on a less costly station, and the data can be moved to a GIS at the end for processing. However, because digitizing is very routine work, operators can become bored and error-prone after long periods of time. Operator duties should be rotated to limit the potential for error.

A typical source document on a digitizer table. (MSE)

Steps of the Digitizing Process

1. **Source preparation.** Data may need to be verified or clarified on the source document. Registration points must be identified, and the document must be secured to the digitizing table.

2. **Map registration.** Four or more locations on the source are matched to the coordinate system and the RMSE (root mean square error) is calculated and evaluated.

3. **Data entry.** The operator follows features and clicks on shape points.

4. **Cleanup.** Digitized data must normally undergo a cleanup process to remove minor topologic errors such as undershoots and overshoots.

Tablet digitizing requires that you orient the map sheet or plat to established ground references and stabilize it on a digitizing tablet surface. Then, by moving the cursor (commonly called a puck) to various strategic points on map features and recording these intercepts with a click, you can identify and register the points, lines, and polygons that comprise map features using a series of spatial coordinates. The figure below illustrates a setup for tablet digitizing. The data file itself is a tabulation of features with identification symbols followed by the string of XY points necessary to create the objects.

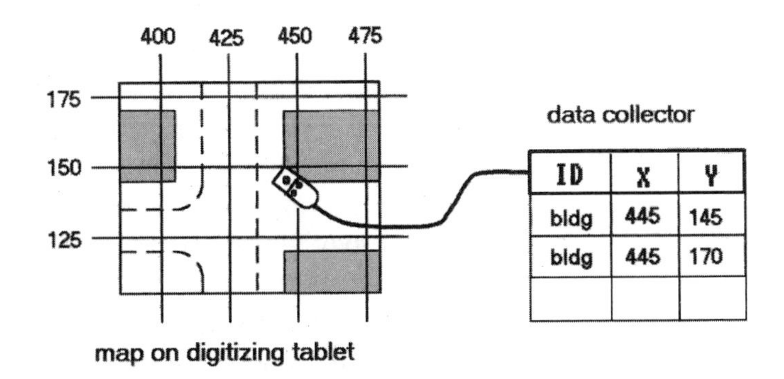

Using a digitizing tablet to digitize existing map detail.

Advantages of Digitizing

- It is inexpensive.
- It is easy to learn.
- Source maps can be captured at nearly identical accuracy.

Disadvantages of Digitizing

- Its accuracy is limited by the stability of the source media and the "drafting quality" of the source.
- It is tedious and therefore error-prone.
- A cleanup step is required.

Using a scanner, you can scan features from maps and plats to form a digital data file of picture elements. However, be aware that these data must be run through vectorizing algorithms before they can be manipulated as mapping data. (For a detailed discussion of this process, see Chapter 13, "Scanning.")

In digitizing and scanning, it may be essential for you to attach topographic information to the horizontal information extracted from the map or plat. Otherwise, the digital data will be two-dimensional and will not work for spatial manipulation. Vertical data may be taken from DEM files and harnessed to the digitized horizontal data, or they may be digitized from a hardcopy line map.

Maps and Drawings

As mentioned previously, the landbase typically consists of cadastral data, planimetric data compiled from aerial photography, or both. In the case of parcel mapping, paper tax maps are typically the source for these data. Depending on your application and accuracy requirements, linework on tax maps may simply be digitized "as is." Current industry practice is to scan documents (if they are stable and sufficiently robust), convert the data from raster to vector format using software, and edit the data interactively (in "post processing"), depending on the accuracy and condition of the source document. (See Chapter 13, "Scanning," for more information.) You may also take steps to "warp" draftsmanship that may have eroded over a document's lifetime to achieve greater spatial accuracy and link the document to established geodetic controls. Often, parcel mapping GIS projects reconcile or "best fit" documents to orthophotographic aerial photography. Data conversion companies have developed or purchased software that can mathematically transform linework. While the approach you take depends on the accuracy of the source document relative to your GIS, the predominant approach is to be as accurate as possible within budget and schedule constraints.

Utility companies of all types usually have their assets mapped on paper maps. All features may be mapped on one "atlas," such as a city map of sanitary sewers or storm drains. In the case of more complicated utility networks (e.g., electric utilities), assets are mapped on several map sets. An electrical utility may have sets of circuit maps, overhead and underground maps, street light maps, and others. If you are converting utility data into a GIS, you would need to refer to all sets as potential sources of data particularly if the level of detail you require only exists in certain large-scale maps. In fact, the driving force behind the GIS may be compiling separate maps sets into a single, central database.

City planning departments typically use numerous maps depicting various boundaries that can be used for a GIS.

Zoning, land use, and general plan maps are only a few sources that can be digitized or scanned and vectorized into the GIS.

Feature attributes can also be captured from paper maps and drawings. All maps discussed previously contain abundant information in annotations regarding features. A parcel's identification number; a utility pole or transformer number; and a water main's size, length, and material can be found on maps. In fact, maps may be the only place such information is stored. If you require these types of data, you can either enter the data manually when you digitize associated features or at a later point.

Tabular Data Sources

Because it is often more expensive to capture attribution than graphic features associated with attributes, you should make every effort to procure attribute data from existing digital sources and minimize the amount that must be entered manually. Do not be surprised if such data are stored in hardcopy format, however. Water utilities often store data about specific features on separate sets of cards or files. Hydrant data may be kept in a card catalog of hydrant cards, service lateral data on service cards, and valve data on valve cards. The data on these cards must be matched with features on inventory maps or atlas sheets. They can be entered during digitizing, or as a clerical function and then bulk loaded against a match key. It is also not uncommon to store feature attributes in more than one place. One utility kept valve information on cards and map sets known as "gate sheets." However, because some information was unique to each source, both were used in the conversion project.

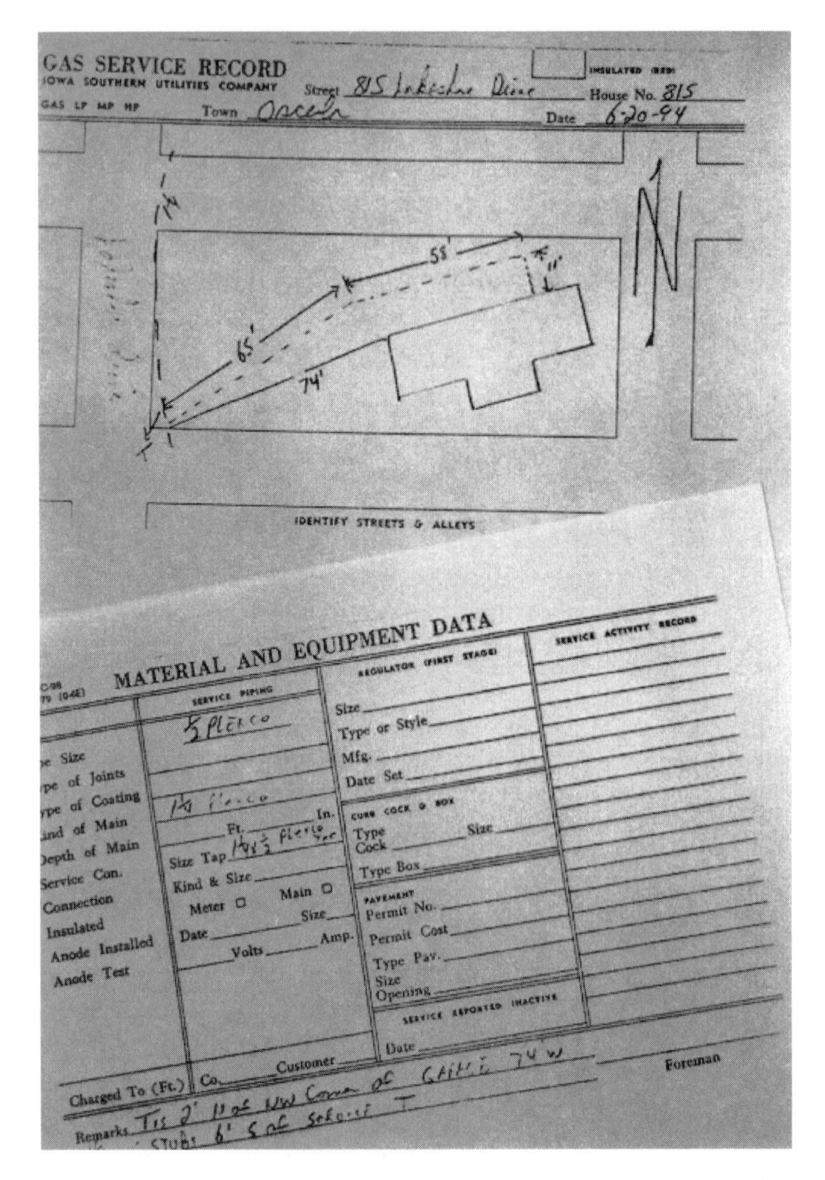
GAS SERVICE RECORD
IOWA SOUTHERN UTILITIES COMPANY
GAS LP MP HP
Street 815 Lakeshore Drive
House No. 815
INSULATED (RED)
Town Osceola
Date 6-20-94

IDENTIFY STREETS & ALLEYS

MATERIAL AND EQUIPMENT DATA

SERVICE PIPING

Size
Type of Joints
Type of Coating
Kind of Main
Depth of Main
Service Con.
Connection
Insulated
Anode Installed
Anode Test

Ft. In.
Size Tap
Kind & Size
Meter ☐ Main ☐
Size
Date
Volts Amp.

REGULATOR (FIRST STAGE)
Size
Type or Style
Mfg.
Date Set

CURB COCK & BOX
Type Cock Size
Type Box

PAVEMENT
Permit No.
Permit Cost
Type Pav.
Size Opening

SERVICE REPORTED INACTIVE
Date

SERVICE ACTIVITY RECORD

Charged To (Ft.) Co. Customer
Foreman
Remarks

Sample service card and data sheet source documents. (MSE)

In some cases, tabular data may also serve as sources for capturing graphic features. Locating legal descriptions of easements for land parcels is one good example. Instead of depicting actual easements on the paper map, only a notation of the deed number and a reference to its location appear. In this case, it may be necessary to use the legal descriptions to enter the extents of the easement using coordinate geometry (COGO).

GIS Data Synthesis

In some cases, the features or attributes you desire may not be available on existing maps or records. Or perhaps only schematic representations of feature locations have been employed to the present. As you implement your GIS, now may be the appropriate time, if your budget and schedule allow, to map the locations more accurately.

Aerial Photography and Planimetric Data Compilation

For some time, engineering companies, utilities, and other entities have relied on aerial photography to perform a wide variety of tasks. Once the science of analytical aerotriangulation matured, the imagery that appears on aerial photographs could be mapped or "traced" to very stringent levels of absolute and horizontal accuracy, in a process known as stereocompilation. Road edges, building and structure outlines, vegetation boundaries—whatever could be seen in the photograph—could be traced on a hardcopy substrate to create a planimetric map. If suitable vertical control was performed, a skilled stereoplotter operator could use the resulting three-dimensional imagery to capture topographic contour lines.

This type of mapping was the industry mainstay until the 1970s. Technology then advanced sufficiently to transfer the tracing function of stereocompilation directly to a computer file, instead of a hardcopy substrate. The computer file stored a series of coordinates that together defined a line or curve and displayed it on screen. Even then the industry made use of hardcopy plots of these files until further advances in software and hardware made automated mapping possible. This technology is now the foundation of GIS, and capturing features from aerial photography often serves as the landbase for GIS applications, with or without parcel information. Indeed, if you know that conventional parcel mapping methods are spatially inaccurate (which is more common than not), it is common practice to redraw and "best fit" parcel linework to features visible on an aerial photograph denoting property lines (e.g., fences and the outside edge of a sidewalk). For further discussion on aerial photography and digital orthophotography, see Chapter 10, "Airborne Sensing Systems and Techniques," and Chapter 11, "Producing GIS Data from Aerial Photos."

Digital Orthophotography

Despite the wealth of information that can be compiled from aerial photography and mapped as vector data, it still can be considered a secondary technology in GIS. Until recently, planimetric mapping was the mainstay of GIS landbase maps because the technology could not support creating, massaging, processing, storing, and displaying imagery in a digital format. Now, hardware and software have evolved such that aerial photography can be processed, stored, and displayed on a GIS workstation. However, because digital orthophotography involves raster data, vector planimetric data continue to be captured for analytical applications that digital orthophotography cannot support.

Digital orthophotography visible on a softcopy terminal. (MSE)

Field Capture of Data

Utilities with GIS environments often find that some degree of field capture of attribute data is necessary because certain attributes have never been cataloged. GIS initiatives certainly provide a window of opportunity to capture such data in the field.

Addresses are notoriously problematic in municipal GIS projects. Historically, compounded inaccuracies "float to the top" when a layer is being built. If this occurs, there is often no choice but to send personnel into the field to perform a site-by-site reconciliation of addresses.

You can also capture features in the field. This process can take several forms. One method becoming increasingly common takes advantage of various global positioning system (GPS) technologies. (For more information, see Chapter 12, "Global Positioning Systems.") Sophisticated *rapid-static kinematic* methodologies can determine positional coordinates in the field (i.e., in real time) or in the post-processing phase. As hand-held GPS receivers become less costly, smaller, and more precise, organizations are relying on them more for positional fixes at specific field locations.

COGO Input

COGO (coordinate geometry) data are the products of field surveys and can be used as a geometric foundation for mapping. In GIS data conversion, COGO is most frequently used to construct very accurate parcel base data. The source data may exist in the form of legal descriptions, records of survey, tract maps, or similar documents. These sources specify the bearing and distance of each traverse leg in the survey. XY coordinates can be determined for each turning point by calculating the latitude and departure from the survey data.

This information is also collected directly from GPS ground receivers by capturing multiple satellite signals and converting them into the 3D spatial coordinates of a survey point.

During traditional field surveys, <XY> coordinates are derived from data collected as horizontal angles and distances in a series of courses through a traverse. The vertical coordinate needed to complete the spatial triplet is gathered as spirit level information.

These data can be input into a GIS file as a basis for creating geographic data. Mathematical instructions in the form of distances and bearings from known survey locations are entered via a keyboard to precisely construct linework far more accurately than most other conversion methods. This accuracy comes at a high price. COGO entry requires very detailed source data, and discrepancies—which are common—must be manually resolved by qualified individuals so that the data fit into the surveyed geographic framework. COGO data cannot be rubbersheeted to fit the survey points because this would destroy the accuracy of the newly entered data.

Optimal Data Source Selection

If you are in charge of orchestrating the database build phase of your project, choose the best source for features and attributes, or the sources that provide the greatest accuracy and are most fiscally prudent. An excellent tool for this purpose is a data/source matrix, which you can create in a simple spreadsheet. First, create a row down the left side of the spreadsheet for each feature and all of its attributes. Next, create a column to identify the feature or attribute's primary source. As you fill in the matrix, you will quickly determine the feasibility of capturing certain types of data. For example, if the only source for sewer laterals in your GIS is through field capture, you may have to delay capturing laterals until it is financially feasible.

Coverage	Feature Class	Attributes	Conversion Strategy	Contractor's Primary Data	Contractor's Secondary Data
Cadastral					
Street centerlines	Arc		Generated from ROW		
		Street name key	Generated and assigned		
Street names	Annotation	Street name/suffix	Key entry from Plan. Map	Planimetric Maps	
Site address	Point		Key entry from Plan. Map	Planimetric Maps	
		Number	Key entry from Plan. Map	Planimetric Maps	
		Street name key	Semiautomate from CTRLINES		Planimetric Maps
	Annotation	Street number	Key entry/cartographic placeme	Planimetric Maps	
Parcels/Condominiums (two distinct coverages)	Polygon				
		Existing PIN	Key entry from Real Estate Maps	Real Estate Maps	
		New PIN	Automated point/X,Y generation		
		Owner name	Bulk load	City-provided digital file	
		"Care of" field	Bulk load	City-provided digital file	
		Owner address	Bulk load	City-provided digital file	
		Owner city	Bulk load	City-provided digital file	
		Owner state	Bulk load	City-provided digital file	
		Owner Zip	Bulk load	City-provided digital file	
		Site address	Bulk load	City-provided digital file	
		Site street name	Bulk load	City-provided digital file	
		Site addr/apt. no.	Bulk load	City-provided digital file	
		Date purchased	Bulk load	City-provided digital file	
		Amount	Bulk load	City-provided digital file	
	Annotation	Existing PIN	Key entry/cartographic placement		
		New PIN	Key entry/cartographic placement		

A section of a data/source matrix spreadsheet.

The matrix should list each attribute, its source, and whether the source is primary or secondary in nature.

A data/source matrix is an excellent tool for confirming that all features and attributes have sources, and it can be a valuable "reality check" on determining whether capturing data is feasible and prudent. If you use a data conversion consultant or firm, the matrix provides the contractor with a wealth of information in summary form. The matrix can also be used to organize and prepare the project's metadata, or data about the data, to answer questions regarding feature origin. Finally, the matrix can be used as the primary tool for tracking the procurement, preparation, conveyance, and return of all source documents to and from an external data conversion vendor.

Feature Quantities

If you plan to use an external data conversion vendor, it is essential that feature quantities are tallied and provided to all prospective bidders. How many poles are there in the project area? How many parcels are within the city limits? Approximately how many polygons representing different soil types are there? These types of questions are inevitable when potential vendors review the matrix, and vendors will be unable to accurately estimate the cost of capturing specific features and attributes unless you provide adequate information. Vendors will understand that you may not have an accurate count for everything and can work with estimates and varying densities of features on a complete set of source documents. Because some vendors charge by the feature, you must have some idea of data quantity.

Data Source Preparation for Entry into the GIS

Once you have determined the project's data sources, you must implement procedures to maximize the quality and efficiency of conversion and data maintenance. The following steps assume that you are employing an external vendor because such situations are usually more complex. In either case, however, adequate source preparation is critical.

Retrieval/Duplication of Data Sources

The maps and records that will serve as sources must be delivered to the data conversion vendor at some point. Traditionally, this is accomplished by making copies of each source document set. The originals may be on diverse media, including Mylar, vellum, or even paper. Experience with these media will determine the best reproduction technology. You may generate copies or you may ask the vendor to complete the task. It is becoming more common for the vendor to scan all source documents, even original ones. By having the vendor scan the sources, a potential source of conflict can be avoided because then the vendor alone is responsible for scan quality.

It is rare for an entire set of source documents to be reproduced and sent to the vendor at once. Typically, the project's scope is divided into phases, and documents are provided in increments corresponding to each phase. This approach also allows you to continue using data for areas that have not been converted. An incremental approach minimizes backlogs in maintaining maps and records that are the result of "freeze dates" (discussed below). Finally, this approach makes it easier for both parties to track the life cycle of a source document.

If your project involves end users distinct from the project team, it will be necessary to generate two copies of source documents. It is critical that both parties have exactly the same source documents when data are received and validated.

Freezing Source Updates

Life goes on during data conversion. Changes to features and attributes will occur at the same time elements are being converted from source documents into a GIS format. It is normally not prudent to update records once you have provided them to the conversion vendor. As such, you must establish a freeze date for each set of maps and records as they are copied and provided to the conversion vendor.

As changes to your data inevitably occur, you must catalog and identify them as changes since the freeze date. Once you receive the GIS data from the vendor, you then update the GIS files, not the manual maps and records. Verify that you identify and track freeze dates for each phase of documents provided to the vendor.

Scrubbing Data Sources

Source documents must often be enhanced before conversion. In a process known as "scrubbing," one or more map attributes are highlighted to separate them from others you do not want to capture. For more information on this process, see Chapter 14, "Keyword Entry of Attribute Data."

Posting of Backlog

Implicit in a GIS implementation is that once your initial conversion is complete and the database is operational, you will henceforth maintain and update features and attributes in the GIS environment while you transition out of manual maintenance methods. This is particularly true because data have been frozen over the course of the conversion. Some degree of backlog occurs, and new information must be "posted" to the GIS. Furthermore, many organizations have an existing backlog when they undertake a conversion project and must address that backlog as well.

Cross-References, Conflict Resolution, and Cleansing Records

No map or record set is error free. The process of data conversion entails "touching" literally every feature and attribute on a source document, and you can be certain that errors contained in paper maps will manifest themselves in your GIS. While you may regret that errors emerge, data conversion offers perhaps the best opportunity to cleanse data and achieve new levels of data accuracy.

Because attribute capture often involves multiple documents, you must be prepared to reconcile conflicts as you go. This process is best handled by individuals familiar with the documents and having "real world" knowledge of what makes sense. The results are then conveyed to the conversion vendor. For more information on this topic, see Chapter 16, "Quality Control/Quality Assurance."

Source Document Inventory, Control, and Tracking

At this point, it should be apparent that you must inventory and track all documents to be converted. A simple spreadsheet can be very effective for this task. In contrast to the data/source matrix, document inventory and tracking is an administrative task. On the spreadsheet, list all sources by type or name, unique map sheet numbers, and dates data are delivered to the conversion vendor (or otherwise enter the process). You may also want to document the tile names/numbers corresponding to particular batches of source documents. The inventory should confirm that the vendor receives all necessary documents, track when the documents are returned (particularly if they are originals), and determine the documents you must use to complete data acceptance testing when the data are delivered.

Acknowledgments

OnWord Press thanks Edgar Falkner and Dennis Morgan for contributing the "Digitizing Process" and "COGO Input" sections of this chapter.

Data Models, Collection Considerations, and Cartographic Issues

Samuel Ngan

GIS data sources can take many forms. This chapter explores raster and vector data types, as well as primary and secondary data collection. Finally, cartographic issues regarding data sources are addressed.

Raster versus Vector Data Representation Issues

All details of a geographic surface cannot be represented in a digital database. Moreover, end users may only be interested in particular subsets of data for a specific geographic area. Therefore, when your GIS database is created, you must explicitly define the kind of information it will store, what types of geographic data to use, and how to structure the information in the database. Generally referred to as data modeling, this process is vital to database creation. It becomes the framework for storing abstract representations of the real world, reduces the volume of data collection, and allows users to choose data sets appropriate to particular interests or problems. To store digital data in a GIS database, two models have been widely adopted: vector and raster. (For more detail on how they differ, see Chapter 2, "Data, the Foundation of GIS.") Since the advent of GIS, industry participants have debated the merits of raster and vector data. These arguments, however interesting, are misguided because advantages and drawbacks are inherent in each model. The following sections address four major issues you should consider when choosing data for GIS data conversion.

Characteristics of Geographic Phenomena

Because the raster data model stores geographic phenomena as a collection of grid cells, the model's basic component is a location in space. All geographic features on the Earth's surface can be *implicitly* represented as appropriate values in cells. In addition, a group of cells representing a geographic feature is encoded with a unique number to identify the cells as part of the feature. Therefore, raster format is commonly used in remote sensing because it analyzes objects from the radiation they emit or reflect; unique identifiers for objects are spectral signatures. Because remote sensing is a major data source for environmental and resource management, the raster format dominates in these fields.

In contrast, the vector model stores geographic phenomena as entities or objects in a database. By defining topological relationships among all objects, network linkage can be *implicitly* described in vector formats, which can facilitate geographic modeling operations, such as network analysis, that require you to perform operations on an object.

Next, because raster data comprise a regularly spaced sampling of phenomena, they reflect little knowledge of spatial variation. For example, a large homogeneous region such as lake is encoded as multiple grid cells having the same value. Obviously, raster representation wastes sampling efforts in less spatially variant areas and reveals little information about spatial variation in complex areas. Therefore, it is inappropriate for applications that require more intense sampling in some areas than others, such as economic and demographic analysis, because it is difficult to vary sampling in response to spatial variation. This problem can be overcome by the vector format, because it allows changes in spatial variability.

Finally, some spatial objects are vector format by definition, such as property lots defined by sets of legal coordinates bounding an area. For this kind of information, data collected in vector format are typically coordinate geometry (COGO) and legal boundaries.

Locational Precision

In raster representations, locational precision is limited by the size of grid cells. Larger grid cells result in more generalized spatial features and poorer locational precision. Locational precision is inversely proportional to grid cell size; the larger the grid size, the poorer the locational precision, and vice versa. For applications requiring more accurate locational precision, such as urban mapping, high resolution orthophoto data should be used, which allows detailed planimetric information to be identified. However, for small scale, statewide mapping projects, low resolution satellite imagery may be acceptable to represent the large extent of area features.

In vector representations, locational precision is limited by the internal coordinate representation used in the GIS system. Typically, GIS can provide up to 16 decimal digits to represent coordinates. For most mapping and surveying applications that require accurate positions of interested objects, existing vector based GIS generally provides sufficient geometric accuracy, so that resolution problems encountered with raster data can be eliminated.

In raster representations, it can be unclear whether the center of the cell or one of its corners is the true location of a coordinate because different GIS systems use varying cell reference points to represent coordinates. Some systems use the top left corner of a cell as the reference point, while others use the middle of the cell. This leads to locational error of up to half the cell's width and height.

Speed of Analytical Processing and Computing

Because spatial relationships are not inherent in the raster model, many geographical analyses (e.g., map overlays, radial searches, and Boolean queries) can be addressed as simple comparisons of neighboring cells. However, because all spatial relationships must be explicitly defined in vector systems, an infinite number of potential spatial relationships is possible depending on the application. This slows response times for queries.

The process of creating polygon overlays illustrates this important distinction. In a raster based system, determining whether and where two polygon features intersect is simple because each polygon is composed of sets of pixels that can be examined to evaluate whether they intersect one another. However, in vector systems this task becomes considerably more difficult because all intersections between the polygons must be computed, which can be time-consuming and tedious.

Data Storage

Because the basic unit of information in the raster data model is a single grid cell, you can improve the resolution of the study area by decreasing the size of the smallest grid cell, and hence capture more information. However, keep in mind that making grid cells smaller increases the volume of data that must be stored. As you may suspect, data volume and grid size are inversely proportional.

The most common and simplest method for storing raster data is to assign each grid cell to a single unit in the database. However, this method wastes storage space if the data cover an area with little spatial variation. To reduce waste, data compression and decompression methods (e.g., run-length code and quadtree) have been developed. The major pitfall of data compression is the extra time it takes to "pack" (compress) and "unpack" (uncompress) data.

In the vector model, data volume depends on the complexity of features in the study area, locational precision, and the spatial relationships defined among features. Generally, when resolution and accuracy are comparable, vector data occupy less storage space than raster data.

Although both data representations have the capability of accommodating any type of geographic data, the advantages of each model depend upon the data you need and applications you plan to use.

The following table summarizes three issues that should govern your choice of data format: the source of data, geographical characteristics, and application requirements. In mapping, surveying, and utility information system applications, where accurate geometric location and accurate representation of the original data are essential, the vector based approach is prevalent. In environmental and resource management applications, where remote sensing is the key data source and continuous geographic variations of data are acceptable, the raster based approach is more common.

	Vector	*Raster*
Characteristics	• Supports topological relationships • Better suited to small amounts of data • Time-intensive data input and management • High degree of geometric accuracy • High quality graphic representation	• Continuous geographical variation • Fast data capture, usually directly from data input device

Data sources	• Manual or automated digitizing from hardcopy maps • Coordinate geometry (COGO) • Third party data (DLG, TIGER/Line) • GPS/surveying • Photogrammetric surveys	• Remote sensing • Digital orthophotographs • Scanners • CCD cameras
Applications	• Planning and emergency applications • Parcel map based information management • Linear network analysis and modeling	• Environmental and resource management • Orthophoto mapping • Terrain modeling

Hybrid Raster and Vector Representation

As computer systems become more powerful, a hybrid format has emerged that integrates raster and vector based data. Many raster based systems on the market now add mapping capabilities to their image processing functions, while vector based systems provide raster and image capabilities. In addition, the number of systems that can convert data between the models is growing substantially. Generally, hybrid systems deliver a bundle of tools allowing users to combine the functionality typical of both data formats.

Because hybrid systems offer greater flexibility, they have been widely adopted for data conversion and public information access. In the data conversion process, operators use raster based scanned images as backdrops for heads-up digitizing, or raster-to-vector conversion utilities to monitor the automated vectorization process. For GIS, hybrid systems are widely used to display vector based cadastral (parcel based) maps over orthophotos, as well as in utility networks.

Because hybrid systems draw on the advantages of both formats, the dilemma of choosing raster or vector is no longer a major issue in GIS database development. Instead, the challenge has shifted to how diverse geographic data can be integrated into a single GIS database. This topic is covered in the following section.

Data Issues

Because the success of a particular GIS greatly depends on the quality of information stored therein, the goal of geographic data collection is to create a clean and usable digital database suitable for your applications. To develop such a database, you must understand three issues relevant to data sources, collection methods, and your target system.

- **Currency.** This concept includes two major components: lineage and age. Lineage describes source materials and how they were acquired (or derived), while age relates to the date source materials were acquired, how they were compiled, and their publication.
- **Precision.** In GIS, precision is considered the number of significant figures in a measurement.
- **Accuracy.** The notion of accuracy is divided into two components, content and position. Position is a measure of how closely the data results, computations, or estimates compare with true values. Content pertains to how complete the information is, its logical consistency, and feature identification.

Data Collection in GIS

There are two tiers of geographic data collection in GIS, primary and secondary. Primary data are collected directly from the field. Secondary data are derived from existing documents or other sources.

Primary Data Collection

Currently, four major sources are considered primary: geodetic surveying, surveying, photogrammetry, and remote sensing. Although they are discussed briefly in this chapter, refer to Section 5, "Data Conversion/Input Methodologies," for further information.

Geodetic Surveying and Geodetic Control Networks

Creating a geodetic control network is an essential step in every primary data acquisition effort. Through geodetic control networks, physical and legal features in a study area can be precisely located relative to their actual positions on the Earth's surface. Understanding how to create a geodetic control network is important because it can influence the accuracy of your data acquisition effort, as well as future database integration and development.

To establish a geodetic control network, include the following elements in your GIS database specifications. A licensed surveyor can help you determine specific survey requirements, tools, standards, and appropriate methologies that should be employed in your conversion.

- The coordinate system used (e.g., U.S. survey feet, international meters)
- The horizontal datum (e.g., NAD 27, NAD 83)
- The vertical datum (e.g., NAVD 29, NAVD 88)
- Survey instruments used (such as dual frequency GPS receivers)
- Accuracy and precision standards used for new control points (e.g., NGS second order class 0 requirements)
- Adjustment methods used (e.g., the least squares method)

Horizontal Datum

A horizontal datum is defined by the latitude and longitude of a point, the azimuth of a line from that point, and the two radii needed to define the geometric reference surface that best approximates the surface of the Earth in the region of the survey.

One well-known datum, the North American Datum of 1927 (NAD 27), uses the measurement of the Earth known as the Clarke 1866 spheroid as its reference spheroid in order to approximate distances along the Earth's surface. In 1986, the National Geodetic Survey (NGS) redefined and adjusted the existing horizontal control data to produce a new reference system, NAD 83. Consistent with global positioning system (GPS) technology, NAD 83 corrects distortions found in NAD 27 and serves as a global "best-fit" horizontal reference datum.

Vertical Datum

A vertical datum ensures that elevation and depth measurements are held to a common vertical standard. For example, the North American Vertical Datum of 1929 (NAVD 29) is a best fit to several mean sea level tide stations in North America. It was superseded by the revised North American Vertical Datum of 1988 (NAVD 88).

Existing controls use NGS controls and high accuracy reference network (HARN) points. A division of National Oceanic and Atmospheric Administration (NOAA), NGS is responsible for establishing, developing, and maintaining the national spatial reference system (NSRS) using advanced geodetic, photogrammetric, and remote sensing techniques. NSRS is a consistent national coordinate system comprised of approximately 1 million survey points, and it serves as a common reference for measurements of latitude, longitude, height, scale, gravity, and orientation throughout the United States. HARNs upgrade NAD 83 networks on a state-by-state basis. They consist of existing NSRS stations and newly established high order classes of stations (usually having A or B order accuracy standards, as defined by the Federal Geodetic Control Subcommittee), with 60 mi or less spacing.

Surveying

Before the advent of photogrammetry and GPS technology, terrestrial surveys were the primary method for topographic and planimetric data collection. Terrestrial surveying currently relies heavily on the use of electronic distance measurement (EDM), which determines locational information by measuring successive horizontal and vertical angles in addition to distance. Because it yields very reliable and accurate data, it remains the primary method for large-scale engineering mapping and legal surveying.

In the past decade, GPS surveying has become one of the most promising survey methods for establishing survey controls and collecting spatial data. For more information regarding how GPS technology works, see Chapter 12, "Global Positioning Systems." Because it is so accurate, GPS is commonly used to collect utility and storm water inventory data. However, neither terrestrial nor GPS surveying is economical for large data collection projects involving volumes of topographic and planimetric data. Instead, photogrammetric methods are used.

Photogrammetry

For several decades, photogrammetry has been used to acquire topographic and planimetric data. Keep in mind, however, that proper planning and acquisition procedures are critical for successful photogrammetric data collection. In addition, with the advent of analytical and digital photogrammetry, digital information compiled from aerial photographs and digital orthophoto imagery have also become a major source of planimetric data. These topics are covered in detail in Chapter 10, "Airborne Sensing Systems and Techniques."

Remote Sensing

Broadly defined, remote sensing includes photogrammetry. Because data capture remains a major bottleneck in the creation of GIS databases, remote sensing is a popular data source. Satellite systems such as LANDSAT and SPOT are capable of acquiring data for a large area in a very short period of time. Furthermore, these data are already in digital form, ready for GIS manipulation. However, it is still necessary to assess key elements of remote sensing data prior to using them in a GIS.

Secondary Data Collection

Secondary data acquisition refers to the process of converting existing maps or other documents into a suitable digital form. This can be accomplished using manual or automatic digitizing methods to acquire data from existing maps, or coordinate geometry (COGO) to enter feature coordinates from legal documents. There is a risk, of course, that existing maps are outdated and require revision or data supplements from primary collection methods. Existing maps may also introduce error into your database for the reasons listed below.

- Maps may have different scales or levels of generalization. Features may be displaced, omitted, simplified, or exaggerated.
- Maps may be on different datum or coordinate systems.
- If you are completing a small-scale map project, your source maps may use one or more different projection systems.
- Not all maps meet National Map Accuracy Standards. (See Chapter 11, "Producing GIS Data from Aerial Photos.")
- Paper, an inherently unstable media, deteriorates quickly as it ages and is exposed to environments that vary in temperature, light, and humidity.

COGO

COGO, which involves calculating and entering feature coordinates into a database, is particularly useful for compiling land record information from legal documents. Because COGO is capable of extremely precise locational information, it is well suited to legal applications. However, COGO can be very time-consuming and far more expensive than manual digitizing—four to 20 times more expensive, in fact. More importantly, the maps and respective legal sources are not error free. In many cases, adjacent parcels entered using COGO are not edge-matched and require substantial time to resolve and clean.

Although data conversion from secondary sources is not always ideal, there are substantial advantages in cost and speed over 100 percent field data collection. As field data collection technology continues to improve, however, secondary collection will continue to provide quality supplementary data.

Cartographic Data Issues

Developing a GIS database is a complex task. To create an effective one, you must understand cartographic design, establish topological relationships for spatial queries and modeling, and ensure that accurate links exist between attributes and graphic elements for query purposes. Because map creation and digital map inventory remain the primary uses of GIS in many fields, a thorough understanding of cartographic design is a crucial and fundamental step. The relationship of GIS cartography to existing manual maps must be carefully considered. While modern GIS applications offer flexible output capabilities, they can only accomplish what the data will support.

Cartographic design for a GIS database is controlled by two major factors: geographic information content and map scale. The relationship between the two determines how map features are represented cartographically.

Geographic Information Content

Determining geographic content is the first step in creating a cartographic database. You must identify the themes of information the database will portray and the geographic area covered. Separating spatial data into layers ensures that specific types of information used in the database do not interfere with each other, and spatial operations such as network analysis can be performed on each type of information.

In many mapping projects, the geographic area defined in the project is often divided into smaller spatial units that typically correspond with manual map sheets (which may also have been used as secondary data). This structure not only allows you full control over how you load and select map sheets, but also facilitates how they are reproduced. It may also become a logical production unit for your data conversion project. Of course, you may prefer to adopt a new grid as well.

Map Scale and Generalization

Map scale is closely related to map sheet size and the level of information represented. Map scale controls the amount of detail that can be shown and the size of map sheets required to cover the study area. It is clear that as scale decreases, the total space available for symbols representing geographic features also decreases. There is, of course, a limit to how small map symbols can become because they become too difficult for map users to read.

It is true that GIS can store accurate positions of spatial features collected from GPS or photogrammetric surveying. However, maps plotted or displayed in a GIS are always at smaller scales than the features they represent, and what they can represent graphically is, again, restricted by map scale. This adjustment process is commonly referred to as *generalization.*

In creating a GIS database, four major generalization operations typically occur.

- Feature selection
- Geometric line simplification
- Combination
- Feature displacement

You must take into account database implications when planning to support these operations, and understanding when, how, and why to use them is necessary to effectively convert the data that will ultimately support them.

Feature Selection

Feature selection, the most common generalization process, is depicted in the next figure. As the map is reduced in scale, minor roads are eliminated from view, and only major highways are displayed. This process is heavily dependent on the feature classes defined in the attribute database, which in turn determine the relative importance of features to selectively omit when the map is reproduced, either on a computer monitor or in hardcopy format.

Road hierarchy structure is used to omit less significant roads from smaller scale displays.

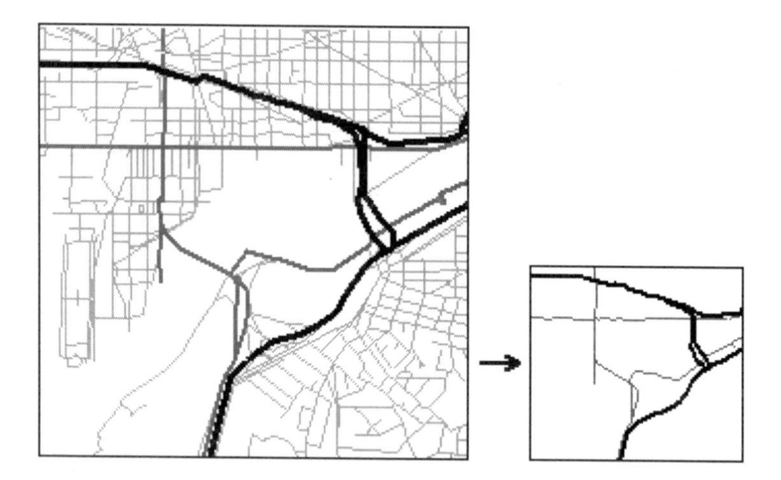

Geometric Line Simplification

When scale decreases, undulating lines are simplified to minimize distracting irregularities. In a process known as stream mode digitizing, the number of vertices recorded is reduced as the process detects coordinates along lines according to time interval or the distance between successive points. Many GIS software packages support line simplification, allowing operators to eliminate redundant vertices, thereby substantially reducing the size of database.

Line generalization retains only major line characteristics.

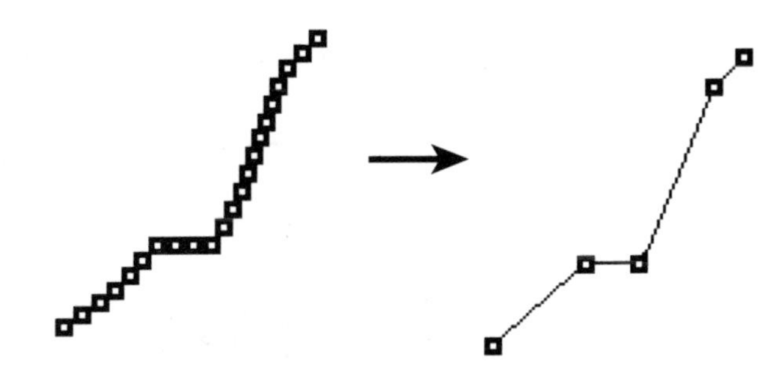

Combination and Displacement

Where many common features exist in close proximity, feature selection is often accompanied by combination and displacement. The following illustration demonstrates this operation. A large-scale map might show details of a service pipe connecting a fire hydrant valve, the hydrant itself, and the distribution main pipe. However, a small-scale map would omit individual features and replace them with an assembly symbol that connects directly to the pipe.

Selective omission and combination.

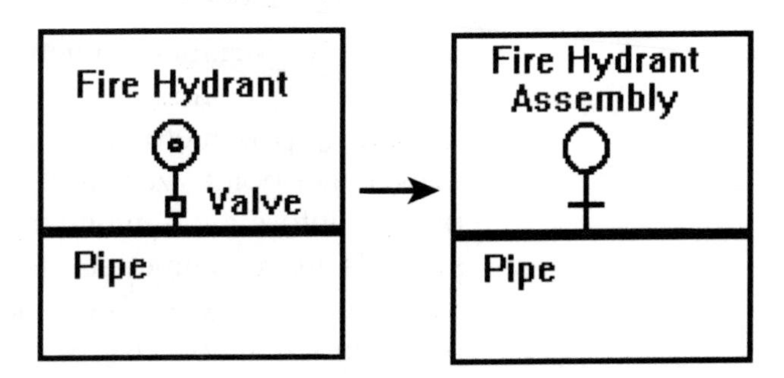

When small differences in position are important to the map user, features will be relocated from their geometrically accurate position (which may be collected from the GPS or photogrammetric surveying) when the map is displayed at a reduced scale. A typical example is water main valves offset from pipe intersections. As you can see in the next figure, relative positions of main valves at pipe intersections may be displaced to differentiate features appropriately at small scales. Geometrically accurate positions for features are retained as coordinates stored in the attribute database.

Feature displacement.

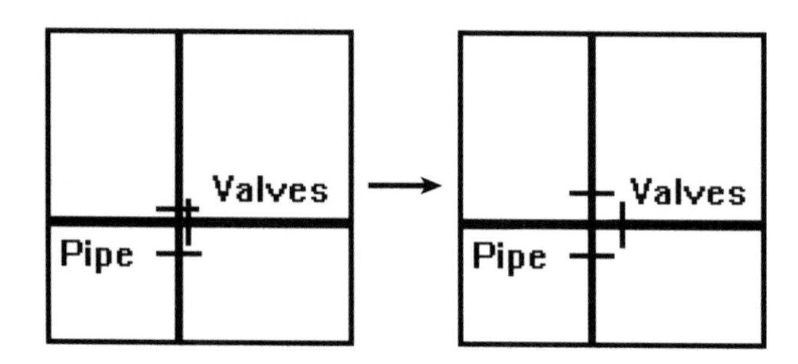

Graphic Representation of Spatial Features

Cartographic databases represent features or phenomena using three types of graphic symbols: points, lines, and areas. Each type of graphic symbol can vary in shape, dimension, and color. The design of each symbol is controlled by the factors listed below.

- **Relative importance.** In general, primary features (e.g., road networks) should be distinct from secondary features (e.g., planimetric information). Therefore, heavier lines, larger point sizes, and stronger colors are used for primary information, while thinner, smaller, and lighter symbols are recommended for secondary information. This makes the symbols distinct and provides pronounced visual contrast for different types of features.
- **Legibility.** Graphic symbols representing spatial features should be large enough to be identifiable under normal viewing conditions. Typically, the minimum dimension for point symbols is 0.2mm, and the minimum gauge for line symbols is 0.15mm. Symbols composed of two or more elements, such as double line symbols (i.e., major roads), should also be separated by at least 0.15mm of space. These guidelines also apply to adjacent symbols.

To allow these symbols to be discernible, ample space (i.e., the minimum gauge) should be created between them in the cartographic database.

- **Recognizable shapes.** Certain map users are familiar with particular symbology. When working with familiar symbols, map users can identify features more quickly and comprehend data more easily. In the past decade, an enormous effort has been put into symbol standardization. One example is the Tri-Service Spatial Data Format, published by the Tri-Service CAD/GIS Technology Center. This format provides a set of standard graphic symbols from which to create the cartography for topographic mapping and facility management applications.

Annotation

Annotation (or map text) is a time-consuming issue in many data conversion projects. It often requires enormous manual effort to monitor the process, and to ensure annotations do not interfere with other map information and are clearly associated with the features they represent. Generally, annotation legibility is controlled by two factors.

- **Choice of type styles and sizes.** In practice, contrasting typefaces or orientations are used to distinguish major feature classes. For example, univers roman typefaces are commonly used for land features, while serif italic faces are chosen for water features. Once the type styles for major feature classes have been selected, you can use different sizes to annotate lesser feature classes according to their relative importance.
- **Annotation arrangements.** The general approach is to place annotations adjacent to, alongside, or within spatial features. In crowded areas, annotations can be stacked, justified differently, and spread out to avoid heavy concentrations of type.

Some GIS software allows users to define a set of rules for annotation placement. In any case, you should take annotation size, font, and frequency into account when you determine how map features will appear when displayed together at different scales.

Summary

This chapter provided an overview of different data structures used for handling geographic data in a GIS database, and covered the major issues of data collection as well as cartographic database development. Inasmuch as good management decisions are based upon the quality of geographic information, the detailed planning of data collection and thorough database design are critical to your GIS. Failure in this area will result in a lack of confidence in your GIS database, and possibly a substantial loss in your investment in the system.

Acknowledgments

I am grateful to Pat Hohl for including me in the book. Without his commitment and assistance, the book could not have been realized. I am also grateful to my former academic advisor, Professor Petrie, for his continuous support since the day I chose the career of mapping/GIS. Finally, special thanks to my family for its encouragement and support; your prayers never failed.

External Data Sources and Formats

Bob Lazar

Considerations When Using External Data

Many government and commercial sources of data exist for use in GIS applications. A number of criteria should be considered when determining the feasibility of using a particular external data source.

Data Formats

Data exist in many formats, such as those specific to particular GIS systems or data producers. Increasingly, formats also comply with formal standards that may be approved by an official standards organization, or de facto standards informally adopted by different vendors. Not all formats are supported in all environments, and different formats may be better suited to transferring some types of data.

Each GIS has its own internal data format. It some cases, this format can be exchanged among different computer systems using the same GIS software package. In other cases, you must create a special export format in the system where the data originate, and then import the export file into the target system. Many vendor formats are proprietary: their exact format is unpublished, and they are not intended for use by other GIS vendors or unauthorized third party vendors. However, some vendors openly publish their formats to encourage data exchange.

Government and commercial data vendors often provide data in formats specific to a particular data product, but standard formats are not specific to systems or data vendors. Typically they are developed and approved by a formal government, industry, national, or international standards organization. Common formats used to transfer GIS data are summarized in the following table.

Format	*Description*
AutoCAD drawing exchange format (DXF)	A vector format that has become the de facto standard for transfer of data between different CAD systems. Often used to transfer geometry into and out of GIS systems, DXF is not well-suited for transferring attribute data.
ARC/INFO export format (e00)	A vector format intended to transfer data, including attributes, between different ESRI systems. Despite being a proprietary format, there are other GIS systems and third party products that can read or write e00 formats.
ArcView shape file format	Openly published, this vector format is available for use by other GIS vendors. It consists of three types of files: main files (SHP), index files (SHX), and dBASE tables (DBF) for storing attributes.
MapInfo interchange format (MIF/MID)	This format, also formally proprietary, has nonetheless been widely implemented in other GIS systems. MIF files store vector graphical information, while MID files store attribute data.
MicroStation design file format (DGN)	An openly documented (except for some newer and product-specific extensions) vector format used by Bentley's MicroStation CAD software. MicroStation is the platform on which the Modular GIS Environment (MGE) and MicroStation Geographics GIS packages are built. The format does not store attribute data, but can store links to relational database records. MGE and Geographics also have export formats that transfer all files and database tables associated with a project.

Digital line graph (DLG) format	A vector format used by the U.S. Geological Survey (USGS), and to a lesser extent, other federal and state government agencies. There are several varieties of the DLG format; DLG optional is most common. This format and USGS data products that use it are often called "DLG-3"; "3" refers to the topological (or 3D) nature of the data. DLG format only supports integer attribute information for spatial objects.
TIGER/Line format	The format used by the Census Bureau to distribute vector and attribute data from its Topologically Integrated Geographic Encoding and Referencing (TIGER) database.
Spatial data transfer standard (SDTS) format	A standard format (used by the USGS, other federal agencies, and in Australia and South Korea) designed to support all types of vector and raster spatial data, as well as attribute data. The Topological Vector Profile (TVP) and the Raster Profile are implementations of subsets of the SDTS. (See Chapter 15, "Spatial Data Transfer Standards," for more information.)
Spatial archive and interchange (SAIF) format	The national spatial data exchange standard in Canada, supporting vector, raster, and attribute data.
National transfer format (NTF)	The national spatial data exchange standard in Great Britain, supporting vector, raster, and attribute data.
Vector product format (VPF)	A vector format with attribute data that is part of the NATO DIGEST standard. It is used by the Digital Chart of the World (DCW).
Tagged image file format (TIFF or TIF)	A raster format frequently used for imagery, GeoTIFF is an extension to TIFF that includes georeferencing information.
Joint photographic experts group (JPEG or JPG) format	Another raster format commonly used for imagery.

Media

In addition to the actual file format, you must also consider the media used to transfer data files because even if you find a useful data source, it may not be delivered on a medium supported by your GIS. For more information on media, see Chapter 6, "Understanding the Target System." Also verify that compression utilities used in the transfer are supported.

Translators

In the simplest transfer scenario, external data would be provided in the internal or exchange format used by your GIS. Using the data may be as simple as copying files onto your system and running the vendor's import command.

However, if data are provided in a different GIS format, you must use a translator program. Many GIS software packages include translators for various external formats, some of which are included with the software at no additional cost. In other cases, the GIS vendor makes translators available for a fee. Third party vendors also sell translation software, either as add-ons or stand alone packages. Free public domain or shareware translators may also be available for some formats. However, free translators may be less robust, less reliable, poorly documented, and unsupported. If no translation software is available, it may be necessary to write your own or hire a consultant to write software to deliver the data in a usable format. If this situation arises, consult your data or software vendor.

Geographic Coverage

Be sure to evaluate the coverage relative to the specific needs of your project.

Scale and Positional Accuracy

Map scales are often assigned to GIS data sets that reflect the scales of hardcopy sources. For example, USGS 1:24,000 DLG files are digitized from 1:24,000 topographic maps. Scale is a vital indicator of data accuracy, and the scale at which is it appropriate to display the data.

Positional accuracy may be expressed by a measurement in paper or ground units. This measurement compares coordinate points in the data set to a more accurate source, such as a geodetic control survey. Such accuracy statements usually indicate that a certain proportion of points in the data set is within a given distance from equivalent points in the more accurate source. The Geospatial Positioning Accuracy Standards developed by the U.S. Federal Geographic Data Committee and the older U.S. National Map Accuracy Standards use this methodology. (For more information, see Chapter 11, "Producing GIS Data from Aerial Photos.")

Timeliness

For all data sets in your conversion project, ensure that you obtain the date when the data were gathered, and evaluate it against your requirements. Double check that this is the date for which the data are current, not the date when the map was printed or digitized.

Coordinate Systems

It is important to know the coordinate system and reference datum the external data use, and what may be necessary to transform the coordinates into the system used in your GIS. Preferably, data should be in latitude and longitude or a coordinate system (map projection or grid system) convertible to latitude and longitude.

The external data format should store information about the coordinate system, and the translation software should be able to read, maintain, and store coordinate system parameters in your GIS. Unfortunately, not all formats store information about the coordinate system being used, and not all translators can transfer this information to your GIS environment. Be aware that you may need to manually enter projection information for certain data sets.

Data Models and Attribute Information

As mentioned in the previous chapter, verify that the data model used (i.e., vector or raster) is consistent with your needs. You must also determine whether your GIS supports all types of spatial objects existing in the external data. If not, does your GIS ignore unsupported spatial objects, or does it convert them to a usable format? For example, CAD formats often contain curved objects. A GIS might ignore these objects or convert them into lines, but the result may not match the original curves.

Determine in advance, if possible, how your GIS stores attribute information as well. Some transfer formats are able to store attribute information about spatial objects, while others can only store linkages from spatial objects to database records. Maintaining links when the data are converted is essential. When translating such formats it may be necessary to transfer attribute data separately using a different format, such as a database export or ASCII format.

Even if a format supports attribute data, a translator may not support all available attribute information. In some cases, graphics may be converted, but attribute data are ignored. Next, a translator may not translate all attribute fields. For example, some translators only support the first pair (out of 20 possible pairs) of attribute codes in the DLG format.

Data Cost

Cost can be a consideration when obtaining external data. Some data, especially government data, may be available for free or at low cost. Other data, while more costly, may provide significant advantages because of their more convenient format, current information, additional attribute information, more accurate geometry, and lower implementation cost.

Examples of External GIS Data

Base Map Data

USGS Digital Line Graph

The USGS has converted many of its printed maps into a digital line graph (DLG) vector format. DLG data contain point, line, and area features separated into categories. Attributes identify specific feature types and additional information. The three major series of DLG data are discussed below.

NOTE: *The term "DLG" is commonly used to refer to vector data produced by the USGS, even if the data are not in one of the DLG file formats.*

Large-scale DLG Data

This series contains data created from the USGS 1:24,000 topographic map series and maps at other large scales (i.e., ranging from 1:20,000 to 1:63,360). Data are available in the following categories: transportation, public land survey system, boundaries, hydrography, hypsography (contours and spot elevations), cultural features, vegetative surface cover, non-vegetative surface cover, and survey markers. Data are available in 7.5' x 7.5' quadrangles corresponding to topographic maps. Nationwide coverage is not complete. All data categories are not available for all quadrangles In some instances, no data are available for certain quadrangles.

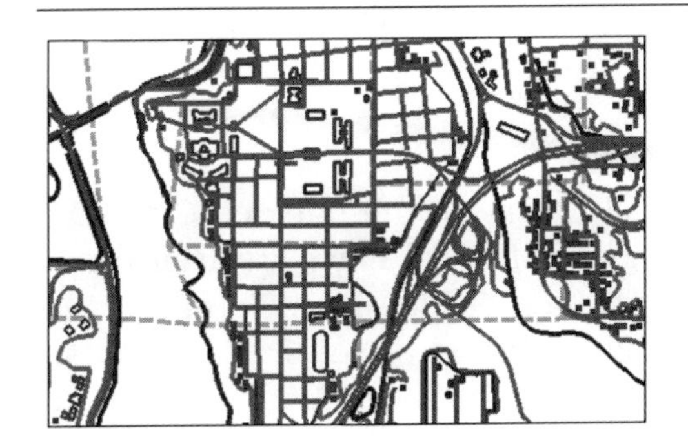

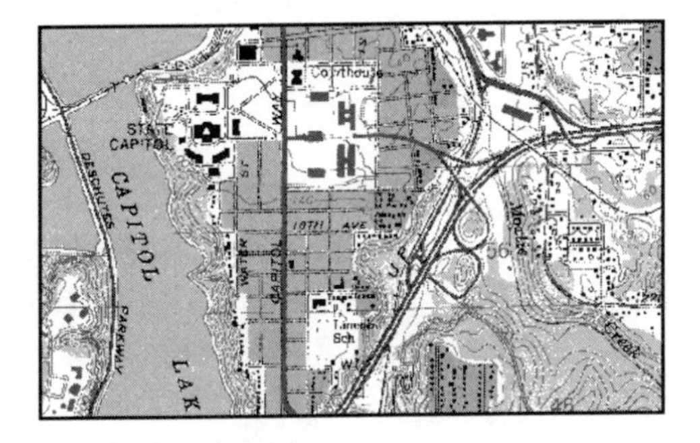

Sample large-scale data: DLG (top left), DLG (top right), and TIGER/Line (right).

Intermediate-scale DLG Data

This series contains data created from the USGS 1:100,000 topographic map series. Data are available for all states except Alaska. Coverage is complete for transportation and hydrography; most of these data sets date from the mid- to late 1980s. Partial coverage is available for hypsography, boundaries, and the public land survey system.

Small-scale DLG Data

This series contains data created from the 1:2,000,000 maps comprising the *National Atlas of the United States*. Two versions of these data are available. The original 1:2,000,000 DLG data date from the late 1970s and early 1980s and partition the United States into 21 sections containing boundary, hydrography, and transportation categories. A revised version of 1:2,000,000 DLG data was produced in the early 1990s. This version is organized into separate files for each state or territory and adds cultural features (i.e., cities) and the public land survey system to the previous categories.

Distribution Methods

The USGS distributes DLG data on the Internet, CD-ROM, and a variety of tape media. Most DLG data are available in either DLG optional or SDTS TVP formats. To encourage acceptance of SDTS, the USGS has placed DLG data at all scales in SDTS TVP format on the Internet where they can be obtained free of charge.

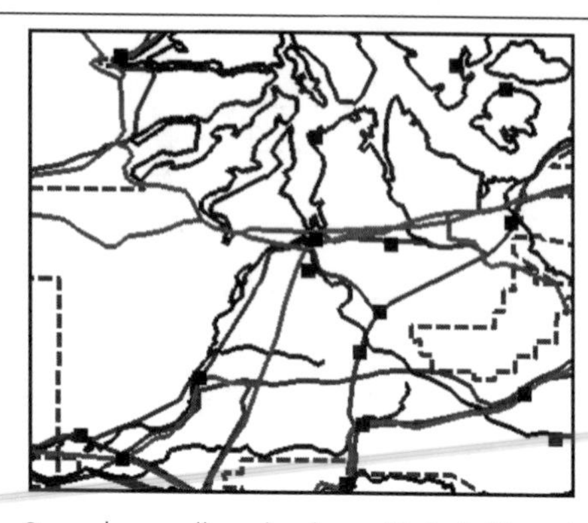

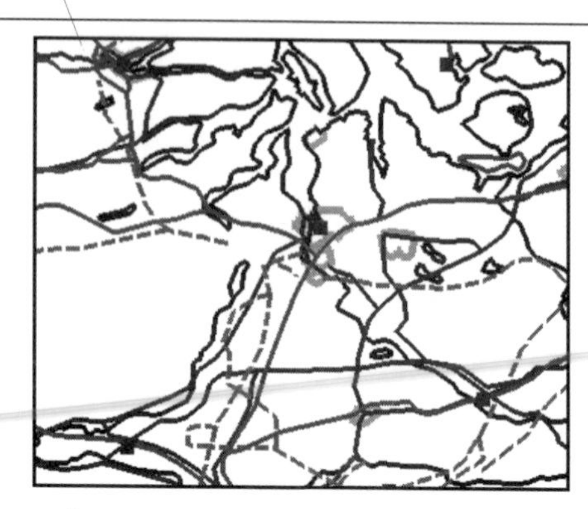

Sample small-scale data: DLG (left) and DCW (right).

➾ **NOTE:** *Small-scale DLG data dating from the late 1970s and early 1980s are still available from the USGS and some commercial and Internet sources. Make sure that you are obtaining the more current state format data, not the older sectional format data. Small-scale DLG data available on the Internet from the USGS are in the more current format.*

USGS Digital Raster Graphics

Digital raster graphic (DRG) is the term used by the USGS for raster data created by scanning images of maps. The USGS has produced DRGs from many of its large-scale topographic maps. The data are in a TIFF format, with GeoTIFF extensions to describe cartographic and georeferencing information. DRGs are sold by the USGS on CD-ROMs containing images for all topographic quadrangles within a 1 x 1° cell.

DRGs have many possible uses in GIS applications, although they lack the topological and attribute intelligence that make vector data such as DLGs and TIGER files (see the following sections) more powerful. Many GIS environments can display raster images such as DRGs as a registered backdrop, providing a base map for other vector data or a source for digitizing new data.

Products similar to DRGs are available from commercial sources, including Land Information Technologies and Horizons Technology. Some federal and state agencies produce products similar to DRGs for specific areas of the country.

USGS Geographic Names

The Geographic Names Information System (GNIS) is the U.S. government's official repository of domestic geographic names of physical and cultural features. It is maintained by the USGS in cooperation with the United States Board on Geographic Names, the group that approves names formally recognized by the federal government. The

GNIS includes most feature names found on USGS 1:24,000 topographic maps, as well as names obtained from other federal, state, and local sources. Information about each name includes the federally recognized name, other names for the feature, elevation, state, county, latitude, and longitude. GNIS data are available from the USGS on the Internet, CD-ROM, and nine-track tape. An on-line Internet search interface is also available.

TIGER/Line Database

The Topologically Integrated Geographic Encoding and Referencing (TIGER) database was developed by the Census Bureau beginning with the 1990 census. The Bureau makes these data available to GIS users through extracts known as TIGER/Line files.

TIGER/Line data cover all states and territories with county based files. The files contain line features such as roads, railroads, hydrography, and political and statistical boundaries. These line features form the boundaries of geographic entities such as states, counties, minor civil divisions, incorporated places, Native American reservations, congressional districts, voting districts, metropolitan areas, census tracts, block groups, and blocks. There are also point and area landmark features, such as schools, churches, parks, and cemeteries. All features are contained in one topologically integrated layer.

Geographic entities are identified through standard Federal Information Processing Standards (FIPS) codes and Census Bureau codes. This process allows the TIGER data to be matched to census and other statistical data provided by the Census Bureau and private companies. Attributes on roads include street names and address ranges. The topological structure and attribute data in TIGER/Line files allow you to match street addresses to geographic entities.

The geometric data for major urban areas in TIGER/Line files derive from 1980 Geographic Base File/Dual Independent Map Encoding (GBF/DIME) data, the predecessor to TIGER. Geometric data for other areas are the same as 1:100,000 scale DLG data from the USGS. The data from GBF/DIME areas in TIGER/Line files are less positionally accurate, have fewer shape points (coordinates that show what curves a line makes between end points), and are less cartographically attractive than areas with data from DLGs. However, the extensive attribute information available in TIGER, its ability to link with attribute data from other sources, and its characteristic integrated topological data set make TIGER the data source of choice for many GIS applications.

TIGER data are sold by the Census Bureau in TIGER/Line format on CD-ROMs by groups of states. The entire country is covered in six CD-ROMs. Some TIGER data may be available unofficially on the Internet.

NOTE: *There are different versions of TIGER/Line data, identified by the year in which they are released. At the time of this writing, the most recent version is TIGER/Line 1995. However, TIGER/Line data for 1997 are scheduled for release in early 1998.*

Digital Chart of the World

Digital Chart of the World (DCW) is the most comprehensive worldwide vector data set for use in GIS systems. It was developed by foreign governments' defense agencies and the Defense Mapping Agency, now part of the National Imagery and Mapping Agency. DCW is based on 1:1,000,000 scale and 1:2,000,000 scale sources. The Chart includes hydrography, land cover, physiography, political boundaries, populated places, transportation, utilities, vegetation, and hypsography. DCW is distributed in vector product format on CD-ROM. In the United States, DCW is sold by the USGS. DCW is also available from GIS vendors and third party data providers, many of whom have converted it to other formats. There are also unofficial sources of DCW on the Internet. The second edition of DCW is called Vector Smart Map Level 0 (VMAP0).

Commercial Vector Data Sets

There are many commercial sources of vector data, most of which are similar to TIGER and contain streets and other features with attributes such as street names, address ranges, and codes for political and statistical areas. In most cases, TIGER is the original data source for such commercial data sets, but a vendor may add substantial value to the data. Commercial data sets may be more current, and contain more positionally accurate geometry, shape points, and streets with address ranges. Commercial vendor products may also contain additional attributes to support applications such as vehicle navigation, or they have been reformatted to facilitate ease of use and data conversion. Companies that produce TIGER-like data sets include Etak, Geographic Data Technology (GDT), Wessex, Road Net Technologies, Navigation Technologies, Claritas, Urban Science, and BLR Data.

Many companies sell demographic, statistical, and economic data that can be used with TIGER and similar data sets. Data sets may be boundary-specific, in that they contain the boundaries of specific types of area geographic entities such as zip codes, census tracts, counties, and minor civil divisions. DLG and DCW data are also resold by private companies, usually in easier to use formats with added value.

Thematic Data Sets

Many specialized data sets are of interest to small groups of users studying specific physical, biological, or social phenomena. Numerous federal agencies have data sets related to their specific functions. A few examples appear in the next table.

Type of data	*Source*
Geologic and hydrologic data	USGS
National Wetlands Inventory data	Fish and Wildlife Service
Flood plain data	Federal Emergency Management Agency
National park data	National Park Service
Shoreline data	National Oceanographic and Atmospheric Agency

The USGS produces land use/land cover (LULC) data that detail urban (or developed) land, agricultural land, rangeland, forest land, water, wetlands, barren land, tundra, and perennial snow or ice. There are also associated data for political units, hydrologic units, census county subdivisions, federal land ownership, and state land ownership. Coverage is available for all states except Alaska. LULC data are available in one of two unique formats: the Geographic Information Retrieval and Analysis System (GIRAS) vector format and the Composite Theme Grid (CTG) raster format. LULC formats are supported by some GIS systems, but not as widely as other formats. LULC data are available on the Internet, CD-ROM, or magnetic tape.

The Bureau of Transportation Statistics (BTS) distributes the National Transportation Atlas Databases (NTAD) on CD-ROM and the Internet. The NTAD includes a data set of the National Highway Planning Network containing 400,000 mi of roads receiving federal aid. Sources for this data set include USGS 1:100,000 scale DLGs and other sources of equal or higher positional accuracy. Transportation-specific attribute information includes road length, number of lanes, route numbers, and pavement type. Similar data sets exist for railway, waterway, transit, and commercial air networks, as well as point data for airports, Amtrak railroad stations, intermodal transfer facilities, and water ports. "Background databases" such as place names, boundaries, and data for transportation networks in Canada and Mexico, also exist. One drawback to NTAD is that it uses its own data format, which is not well supported by GIS software packages. However, BTS does provide a viewer with limited translation capabilities.

Some data sets may be developed for specific projects or studies, such as the Upper Mississippi and Lower Missouri Data Base, which was developed in response to the great flood of 1993. These data, available from the USGS, contain a variety of data types collected from various sources.

Elevation Data

Elevation data can be used in GIS software to generate shaded relief and contour maps, create perspective views, analyze visibility, and calculate slope. Elevation data are generally provided in a regular grid of elevation values referred to as digital elevation models, digital terrain models, or digital terrain elevation data. These data sets are most often produced through photogrammetric techniques or from contours. Elevation data can also be found as vector contours in DLG, DCW, and other data sets.

U.S. Elevation Data

The USGS produces digital elevation models (DEMs) with coverage corresponding to USGS quadrangle maps. DEM data for the standard 7.5' quadrangles are characterized by 30m spacing between elevations. DEM data for 7.5' quadrangles in Alaska have an elevation spacing of two arc seconds of latitude and three arc seconds of longitude. Thirty-minute DEM data cover half of a USGS 1:100,000 scale 30 x 60' quadrangle, with an elevation spacing of two arc seconds.

One-degree DEM data correspond to 1 x 2° (1:250,000 scale) USGS quadrangles. Produced by NIMA, these data have an elevation spacing of three arc seconds. One-degree DEM data are available from USGS on the Internet.

Sample perspective view generated from the Mt. Baker, Washington, 30' (or two arc second) DEM produced by the USGS.

Worldwide Elevation Data

NIMA makes Digital Terrain Elevation Data (DTED) Level 0 for much of the world available to the public. These data have an elevation spacing of 30 arc seconds and are organized in 1° quadrangles. Data are available on CD-ROM and via the Internet.

The USGS EROS Data Center has led an effort to produce GTOPO 30, a worldwide data set that uses DTED Level 0 data for much of the world, but supplements it with data from USGS DEMs, DCW contours, and other sources. Like DTED Level 0, GTOPO 30 has an elevation spacing of 30 arc seconds. GTOPO 30 is available on the Internet.

Shaded relief created from GTOPO 30 elevation data. (RMMC)

The National Oceanographic and Atmospheric Agency's National Geophysical Data Center has produced ETOPO 5, a worldwide data set with 5' elevation spacing. ETOPO 5 includes bathymetric data for ocean areas in addition to elevation data for land areas. ETOPO 5 is available on the Internet and with other data on the Global Relief CD.

Sources of External Data

External data are available from numerous government, academic, and commercial organizations. A significant amount of data is available on the Internet free of charge. Other data sources may be more costly but offer significant advantages. Detective work on your part may be necessary to locate the most appropriate source of external data for your particular GIS needs.

Using the Internet to Locate Data

The Internet is a powerful tool to locate, evaluate, and obtain data. Methods of finding data include using the federal government's National Geospatial Data Clearinghouse, visiting Web home pages of known data producers, and using World Wide Web pages containing compilations of links to data sources.

Free Data Available on the Internet

Many federal, state, and local government agencies have placed data sets on the Internet so that they may be downloaded for free. Perhaps the most significant on-line data source is the USGS. The USGS has placed 1° and 7.5' DEM data, all scales of DLG data, LULC data, and GNIS data on the Internet. The USGS and Microsoft Corporation have formed a partnership that will further expand the amount of USGS data available on the Internet, including all USGS digital orthophoto quadrangles.

NOTE: *Some on-line DEM and DLG data are available only in the relatively new SDTS format. However, SDTS is not yet supported by all GIS software packages. You must use an SDTS translator for SDTS formatted data. DLG and DEM translators will translate only DLG and DEM formats, not DLG and DEM data in SDTS format.*

Although some data sets, such as TIGER and DCW, are not officially available via the Internet, it may be possible to find parts of these data sets on line. Because federal government data sets are in the public domain, users are free to redistribute them, which includes placing data files on the Internet. Researchers and other data users have posted public domain data on numerous Internet sites. In some cases the data are in the original format, and in others they have been converted. These unofficial sites may provide an easy and free method of obtaining data, but exercise caution. The available data may not cover your area of interest, or they may be an older version than data currently being distributed by the agency that produced them.

Examples of Free Data on the Internet

The following table contains a sample list of Internet uniform resource locators (URLs) containing free data.

URL	Description
http://edcwww.cr.usgs.gove/doc/edchome/ndcdb/ndcdb.html	USGS on-line data
http://164.214.2.59/geospatial/geospatial.html	DTED Level 0 data from NIMA
http://www.ngdc.noaa.gov/mgg/global/seltopo.html	ETOPO 5 elevation data
http://edcwww.cr.usgs.gov/landdaac/gtopo30/gtopo30.html	GTOPO 30 elevation data
http://mapindex.nos.noaa.gov	NOAA Map Finder
http://www.nwi.fws.gov/data.html	FWS National Wetlands Inventory data
http://edcwww2.cr.usgs.gov/sast-home.html	Upper Mississippi and Lower Missouri database
http://www.bts.gov/programs/gis/ntatlas	National Transportation Atlas Database
http://govdoc.ucdavis.edu/tiger/tiger.html	TIGER/Line data for California
ftp://spectrum.xerox.com/pub/map	Miscellaneous contributions of public domain data
http://magic.lib.uconn.edu	Connecticut Map and Geographic Information Center
http://www.tnris.state.tx.us/ftparea.html	Texas Natural Resources Information System
http://www.martin.fl.us/GOVT/dept/isd/data_dict	Data for Martin County, Florida

Using the National Geospatial Data Clearinghouse

The Federal Geographic Data Committee (FGDC) is responsible for coordinating spatial data collection activities among different levels of governmental and non-governmental organizations. A primary goal of this initiative, the umbrella title for which is the National Spatial Data Infrastructure (NSDI), is to encourage data sharing among organizations and reduce expensive and redundant efforts. The National Geospatial Data Clearinghouse is the NSDI arm designed to allow data users to conduct on-line searches for data that meet their requirements.

Federal agencies are required to document their data using the Content Standards for Digital Geospatial Metadata, developed by the FGDC and often referred to as the FGDC metadata standard. Metadata, or data about data, describe a particular data set. Metadata typically include the geographic region covered; format, scale, coordinate system, and themes; creation date; source materials used to create the data set; how to obtain the data; and the transfer medium for the data set. Metadata provide a means of evaluating data before purchasing or obtaining them.

The clearinghouse is a decentralized set of servers on the Internet containing metadata that comply with the metadata standard. These servers, primarily established by federal and state government agencies, generally contain non-commercial data. You can search the servers using several search interfaces to locate data by geographic area, date, theme, scale, and other metadata.

The FGDC Web home page is a good starting point for locating existing sources of spatial data. There are links to the clearinghouse search interfaces, information about FGDC activities, and links to federal and state agencies and universities that are sources of information and data. Links to government data in other countries are available, but limited commercial data source links are included.

Link Compilations

In addition to the FGDC Internet site, many other sites have compiled lists of links to sources of spatial data. Most lists include federal, state, local, academic, and commercial sources. The most comprehensive lists are often created by academic researchers. However, government agencies and commercial GIS companies may have valuable links as well.

Internet Starting Points

The following table lists URLs for sites where you can initiate Internet data searches. Included are major government mapping organization sites and sites containing a plethora of links to other helpful spots.

URL	Description
http://www.fgdc.gov	FGDC home page. Links to National Geospatial Data Clearinghouse nodes, federal agencies, state coordination agencies, and other data sources; information on FGDC coordination efforts.
http://mapping.usgs.gov	USGS National Mapping Information. Links to information on USGS data and data available on line.
http://www.census.gov	Census Bureau home page. Information on TIGER and other census data and programs.
http://www.nima.mil	Home page of the National Imagery and Mapping Agency.
http://www.cgrer.uiowa.edu/servers/servers_geodata.html	GeoData Information Sources. Perhaps the best set of links to Internet sites with on-line data and information about data.
http://www.cast.uark.edu/local/hunt/index.html	Starting the Hunt: Guide to On-line and Mostly Free U.S. Geospatial and Attribute Data. Another good set of links.
http://ilm425.nlh.no/gis/dcw/dcw.html	Digital Chart of the World and Data Quality Project. Information on DCW, including links to some free Internet sources of DCW for specific parts of the world.
http://library.berkeley.edu/UCBGIS/gisdata.html	GIS Data Resource Directories. More links.
http://earth1.epa.gov:80/oppe/spatial.html	GIS spatial data site links from the U.S. Environmental Protection Agency.
http://www.esri.com/datahound	ESRI Datahound. Search engine that finds free data on the Internet.
http://www.esri.com/base/common/jumpstation/jump_dom.html	GIS Jumpstation. Links from ESRI, including many federal, state, and local government agencies.

URL	Description
http://www.ccrs.nrcan.gc.ca/linc/index.html	Geomatics Canada. Good starting place for Canadian information.
http://www.ordsvy.gov.uk	Ordnance Survey. The United Kingdom's national mapping agency.
http://www.auslig.gov.au/welcome.htm	Australian Surveying and Land Information Group. Good starting point for information from "down under."

Using Data Producer Web Sites

Once they have been located, data producer Web sites can provide extensive information regarding available data. In some cases, the information is similar to what you might read in hardcopy promotional materials, with general descriptions of the data, geographic coverage information, illustrations, pricing, and ordering procedures. In other cases, the information is much more detailed, taking advantage of the Internet's unique capabilities.

Some sites provide detailed metadata information and other detailed user documentation on line. Query interfaces allow searching for data based on geographic coverage, data set theme, date, scale, and other metadata. Often the search process is enhanced with "clickable" maps, where you can select an area of interest. Sample data may be available for download to preview. You may be able to order or download data directly via the Internet. Demonstration and utility software may also be available.

Examples of Data Producer Web Sites

USGS Web Sites

Perhaps the easiest way to access the various USGS Web sites is via the mapping home page, which contains links to on-line data, information about data, information about hardcopy maps and reports, and links to different parts of the USGS. Because many USGS data products do not yet have nationwide coverage, there are status graphics showing the areas where data are available. In some cases (e.g., DRGs), graphics show the areas in production or planned for production. There are links to on-line metadata, standards, and user guides for USGS products.

One method of searching for USGS data is the Global Land Information System (GLIS), an interactive query system that lets users view abstracts, descriptions, and pricing information for USGS data. A query capability allows users to select data based on geographic coverage (latitude and longitude, place name, state, and map name), date, feature type, and theme. Query results show the map name and other information about the data set. Users can select a data set and add it to a "shopping basket" to place an on-line order. An estimated price is displayed, but you still must speak with a customer service representative to finalize the order and arrange payment.

Connecticut Map and Geographic Information Center Site

This site, built by the University of Connecticut Libraries, illustrates a simple yet powerful interface for locating and selecting state, county, town, and quadrangle based data for the state of Connecticut. Counties, towns, and quadrangles can be selected using a map or by name. Available data sets for the particular geographic area are displayed, and detailed metadata are available. All data sets are on line and can be downloaded at no charge. Many of the data are from federal sources including TIGER/Line data and USGS DLG, DRG, and digital orthophoto quadrangles.

SPOT IMAGE, S.A. Site

The Web site for the SPOT satellite provides information on products, prices, and formats. There are index maps, status graphics, sample images, and sample data that can be downloaded. The Web site allows querying a computerized catalog of SPOT's archived imagery. The user specifies latitude and longitude values for the area of interest, spectral mode (color, or black and white), maximum cloud cover, dates of acquisition, and acceptable technical quality. There is a registration fee for using the search interface, refundable with an order. Guest access with more limited query capabilities is also available.

Examples of Commercial Data Sources

The next table lists contact information for many popular commercial data vendors.

Company	Phone	Internet home page
American Digital Cartography, Inc.	800-236-7973	www.adci.com
BLR Data Corporation	800-316-2572	www.blrdata.com
Claritas, Inc.	800-234-5973	www.claritas.com
Cypress Geo-Resources, Inc	619-549-7046	www.cyp.com
Datamocracy, Inc.	800-382-7802	www.datamocracy.com
EarthWatch, Inc.	800-496-1225	www.digitalglobe.com
National Decision Systems, Inc.	800-866-6510	www.natdecsys.com
Etak, Inc.	800-765-0555	www.etak.com
Geographic Data Technology, Inc.	800-331-7881 x1101	www.geographic.com
Horizons Technology, Inc.	800-828-3808	www.horizons.com/suremaps
Land Info International	303-369-6800	www.landinfo.com
MaconUSA	617-254-2295	www.maconusa.com
Micro Map & CAD	303-988-4940	www.sni.net/micromap
Space Imaging EOSAT	800-425-2997	www.spaceimage.com
SPOT IMAGE	800-275-7768	www.spot.com
Wessex Inc.	800-892-6906	www.wessex.com

Other Sources of Information

Federal Agencies

The USGS operates the Earth Science Information Center (ESIC) to handle phone, mail, e-mail, and in-person inquiries and sales. ESIC sells USGS data, hardcopy maps, and other publications, as well as some products (such as DCW), produced by other federal agencies. ESIC is a source of hardcopy information on USGS data, including fact sheets, status graphics, and price information. The number for phone inquiries is 1-800-USA-MAPS.

TIGER/Line data are sold by the Census Bureau. There is currently no method of officially obtaining TIGER/Line data via the Internet. The customer service phone number for information on TIGER/Line and other data from the Census Bureau is 301-457-4100.

State and Local Coordination Agencies

Many states have agencies that coordinate state government mapping activities and provide information on data available from the government and other sources. Examples include the Wisconsin State Cartographer's Office, the Minnesota Land Management Information Center, and the Maine Office of GIS. The FGDC Web site contains links to the Web sites of many of these organizations.

In some cases, regional or county agencies coordinate mapping activities and may be sources of information and data. In Wisconsin, the Milwaukee County Automated Mapping and Land Information System, a consortium of regional, county, and municipal governments and utilities, share the cost of producing base mapping data. If no central coordinating group exists for your area of interest, you can contact individual government agencies or utilities to determine what data are available.

Be aware that policies regarding data distribution vary among government organizations and agencies. Some support sharing data freely, while others charge nominal fees to recover the cost of distributing data or higher fees to recover part of the cost of producing and maintaining the data. In some cases, you may be asked to sign a licensing agreement that restricts how you use and redistribute the data.

GIS Vendors

GIS vendors can often help you evaluate whether data are suitable for use with their systems. Many GIS software manufacturers bundle data with their software packages, sell data, or have arrangements with data vendors to provide data in compatible formats. Vendors may also sell data conversion utilities, refer you to third party developers, or direct to you to free software for this purpose.

Vendor Internet sites often contain information about data compatible with their systems and links to appropriate data vendors. Some GIS vendors sell data on line or provide information regarding free sources of data. Such information may be incomplete, however, because vendors may be reluctant to publicize data designed for competing products.

Publications

Geo Directory, published annually by *GIS World*, includes a "Data Sources" volume available as a printed book or on CD-ROM. GIS industry magazines such as *GIS World*, *Geo Info Systems*, *Business Geographics*, and publications aimed at specific GIS systems contain advertisements and news items from data vendors.

Acknowledgments

I would like to thank Applied Geographics, Inc. for use of GIS software and the U.S. Geological Survey's Rocky Mountain Mapping Center and EROS Data Center (EDC) Distributed Active Archive Center for permission to use illustrations. The EDC provided an image created from GTOPO 30 elevation data. GTOPO 30 was developed through a collaborative between EDC, the National Aeronautics and Space Administration, the United Nations Environment Programme/Global

Resource Information Database, the U.S. Agency for International Development, the Instituto Nacional de Estadística Geográfica e Informática of Mexico, the Geographical Survey Institute of Japan, Manaaki Whenua Landcare Research of New Zealand, and the Scientific Committee on Antarctic Research.

Section 5: Data Conversion/ Input Methodologies

This section provides detailed information regarding a variety of data input approaches.

- **Chapters 10** and **11** discuss data derived from airborne sensing systems and how they are converted for use in GIS systems.
- **Chapter 12** provides an overview of global positioning systems.
- **Chapter 13** discusses the technology of scanning at length.
- **Chapter 14** offers guidelines for keyboard entry of attribute data that accompany feature data in a GIS.

- **Chapter 15** explains the increasing implementation of spatial data transfer standards, the goal of which is to convert spatial data in a variety of formats among diverse GIS systems without losing information.

Airborne Sensing Systems and Techniques

Edgar Falkner and Dennis Morgan

Data Types

Aerial mapping is a technical science that provides precision maps for many purposes—engineering projects, land development, utility control, GIS layers, natural resource management or exploration, and countless others. Generally, it is the fastest and least expensive procedure for producing a detailed base map when no prior data exist.

Remote sensing devices can scan the ground and collect identical wavelengths (visible and near infrared) and in the thermal (emitted heat) and radar (microwave) ranges. Airborne sensor systems are defined here as remote sensing systems mounted in aircraft platforms. The sensors receive and may transmit electromagnetic energy from or to the Earth in the spectral (from visible light through near infrared), thermal, and microwave ranges. Airborne systems have several advantages. The ability to develop a platform and sensors for a specific purpose is very beneficial, because airborne sensor systems can be equipped with multiple sensors set for specific purposes, and data can be obtained at very specific time periods. Projects can be planned to optimize the system for a specific project, allowing for multiple flight altitudes and sensor bands and variable resolution. However, while the cost of such systems is lower than the cost of satellite systems, for a specific project it can be relatively high.

Aerial photography and other remote sensors rely on capturing data in some portion of the electromagnetic spectrum. The following figure shows the range covered by most sensors discussed in this chapter, which are sensitive to visible light, near infrared reflectance, or thermal emission.

Portions of the electromagnetic spectrum that aerial film or remote sensors may register. (RM)

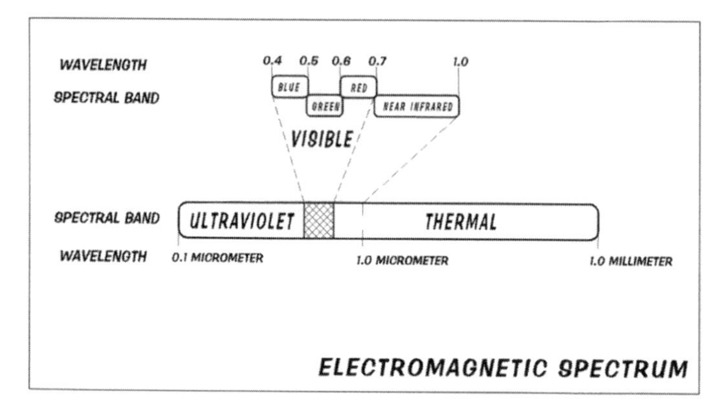

In general, the visible spectrum is comprised of light reflectance in bands 0.4 micrometers (μm) to 0.7μm. Primary colors within the visible range are blue (0.4 to 0.5μm), green (0.5 to 0.6μm), and red (0.6 to 0.7μm). All other colors are formed by mixing primary colors to varying degrees.

In this chapter, discussions of the heat used to compose images refer to reflected and emitted heat. Near infrared (0.7 to 1.2μm) senses the heat reflected from ground objects, while thermal wavelengths sense the temperatures emitted from ground objects. The thermal range is separated into three bands: shortwave infrared (1.2 to 3.0μm), midwave infrared (3.0 to 5.0μm), and longwave infrared (5.0 to 14.0μm).

One of the most common types of sensor data is passive energy data (radiant energy reflected from a source object) in the near infrared portion of the electromagnetic spectrum. These data are commonly used to detect plant health, water boundaries, or particulate pollution. The most common airborne systems for near infrared data acquisition are videography and conventional aerial cameras. Near infrared film is available in either color (false color) or black and white formats, but color infrared, in spite of its higher cost, is more common because of its finer grain, higher resolution, and better interpretive quality. Most aerial photography firms not only own aerial cameras but also have access to high resolution video cameras with the capability to use infrared magnetic tape. Videography systems can "time tag" processed tape for georeferencing and "grab" frames for individual scene processing. Infrared videography and aerial photography both require relatively clear skies with no cloud cover in order to capture ground images. Both systems are common tools for photogrammetry firms and should be cost effective and reliable when used properly.

Airborne thermal infrared sensor systems are also available. These sensors, which are passive devices that record emitted heat from an object in the thermal range of the electromagnetic spectrum, can differentiate actual temperature gradations within a scene.

Airborne thermal infrared data are usually collected at night to maximize recording of specific energy sources. Traditional aerial photography of the site is usually flown prior to or immediately after the thermal flight and is used as an interpretation tool for the thermal data. Thermal sensors are constantly improving and can be expected to record emitted heat energy from a significant depth below the surface. However, thermal data also require significant interpretation, processing experience, and ideally, field checking when the project warrants. Typical uses of airborne thermal scanning include building heat loss, utility system losses (above or below the Earth's surface), and fluid seepage (e.g., hazardous waste or ground water).

Airborne radar systems are becoming increasingly popular. Radar is an active sensor system that transmits energy pulses in wavelengths of 0.8 to 30.0cm that bounce off objects and upon return are captured by the system. Distance between the sensor and object is measured by the time that it takes for the waves to travel to and return from the intercepted ground feature. Radar sensors generally are of two types, synthetic aperture radar (SAR) and side looking airborne radar (SLAR).

Simply stated, SAR—the more common system—creates a synthetically long antenna that can improve the resolution of an image, while SLAR systems feature one antenna length. Because radar can penetrate mist, clouds, and fog, it allows for data acquisition during weather unsuitable for conventional aerial film. Typical airborne radar system uses include high resolution terrain mapping, plant life monitoring, and weather forecasting. Several ground conditions affect radar image data and the ability to accurately depict true ground conditions (including surface roughness and texture). Conventional aerial photography is used to interpret and process final radar data sets. Experienced personnel should be consulted to properly process radar data and account for common radar effects (e.g., shadows and distortion when signals are recorded near the same time). The next table lists several sources of airborne sensor data.

Sources of airborne remote sensing data

*Band range** *p b g r n t*	*Resolution (meters)*	*Agency***	*Comments*
X X X X - -	0.25 - 4.0	PS	1991-97 data
- - - - - X	- -	NASA	75 channels
X X X X - -	0.70 - 3.3	NASA	1987-97 data, 75 channels

Band range* p b g r n t	Resolution (meters)	Agency**	Comments
X X X X - -	20.0	NASA	224 channels
X X X X - -	0.50	NRL	1994-97 data, 206 channels
X X X X X X	1.0 - 10.0	ERIM	1972-97, 31 channels
X X X X X X	12.5	NASA	8 channels

* The letter "p" represents the panchromatic electromagnetic sensitivity band; b, blue; g, green; r, red; n, near infrared; and t, thermal.

** PS represents positive systems; NASA, National Space & Aeronautical Administration; ERIM, Environmental Research Institute of Michigan; and NRL, Naval Research Laboratory.

Types of Photogrammetry

Historically, photogrammetry has been defined as the process of creating maps from photographic images. There are several types of photogrammetry, although the aerial procedure is by far the most common.

Aerial photogrammetry uses images exposed from a moving platform at a distant point in the sky. Terrestrial photogrammetry involves photos exposed from fixed stations at distant points on the ground. Close range photogrammetry relies upon photos from fixed positions in close proximity to the subject matter (often only a few feet away), either suspended above or to the side of the subject.

Advantages of photogrammetry are listed below.

- High level of accuracy (i.e., it carries legal weight).
- Employs a widely accepted methodology that, when properly used, will achieve a predetermined accuracy.
- Image basis (e.g., photographs) can be used as a quality check for mapping completeness.
- Performed by highly skilled technicians.
- Equipment is very accurate and formally calibrated.
- End products can take several forms (e.g., orthophotos, topographic/planimetric maps, and others).

Disadvantages of photogrammetry follow.

- High cost of photogrammetric equipment (e.g., airplanes, sophisticated cameras, and stereoplotters) can be prohibitive.
- Personnel costs can be prohibitive, involving technicians, compilers, photogrammetrists, and geodisists or surveyors for ground control.
- Suitable aerial photography is limited to certain times of the year, depending on the study area, resulting in increased costs and project delays.

Map/Photo Scale

In this text, the term "scale" applies equally to maps, photographs, and sensor images. Scale simply alludes to the relationship of the distance covered on the ground to the same distance measured on the map, photo, or scanned image.

There are two basic ways to express scale, representative fraction, and engineers' scale. Representative fraction is denoted as a ratio (1:1,200), where a single unit on the map or photo equals 1,200 of the same units on the ground, whether measured in inches, feet, or meters. Engineers' scale is expressed as 1"=100', where a single inch on the map or photo equals 100' on the ground. The following table shows the conversion between common scale expressions.

Relative scale expression

Engineers' scale	*Representative fraction*
1"=40'	1:480
1"=50'	1:600
1"=100'	1:1,200
1"=200'	1:2,400
1"=500'	1:6,000
1"=1,000'	1:12,000
1"=2,000'	1:24,000

When discussing scales of photos or maps, two terms are employed, large scale and small scale. This is a relative concept. An engineer designing a subdivision may consider a map to scale 1"=50' as large scale, yet a soil scientist producing a map of soil types for a given county may think of a map to scale 1"=1,320' as large scale. However you think of relative scales, remember that the larger the denominator of the scale expression, the smaller the scale. For example, when comparing photo scales, 1"=200' is considered large scale while 1"=2,000' is considered small scale. Expressed another way, the larger the cultural detail, the larger the scale.

Scale of an aerial photo is a function of the focal length of the camera and the altitude of the aircraft above mean ground level. With this in mind, the only variable in most mapping photo scales is the height of the aircraft. Calculation of photo scale is expedited by the simple formula

$$s = h/f,$$

where s = scale denominator (feet),

f = focal length (inches), and

h = height of aircraft above mean terrain level (feet).

Using the same variables, calculating flight height for an anticipated photo scale would employ the following formula:

$$h = s \times f$$

For example, if a project requires a photo scale of 1"=500' with a 6" focal length camera, then the flight height (above mean terrain level) will be

$$h = s \times f = 500 \times 6 = 3{,}000'.$$

The table below tabulates several photo scales relative to flight heights using a 6" focal length camera.

Photo scale relative to flight height

Scale	*Height*
1"=200'	1,200'
1"=400'	2,400'
1"=800'	4,800'
1"=1,000'	6,000'
1"=2,000'	12,000'

Resolution

Silver grains in photo emulsions and pixels in scanned data both register the same effect: the intensity of light reflected off an object. Yet, the term resolution has two connotations in the photography/GIS arena. Resolution of digital data refers to the dimensions of individual picture elements in the database. Photo image resolution is based on the size of silver particles in the emulsion. Pixels are constant in size and shape, while silver grains are not. The following figure illustrates the difference.

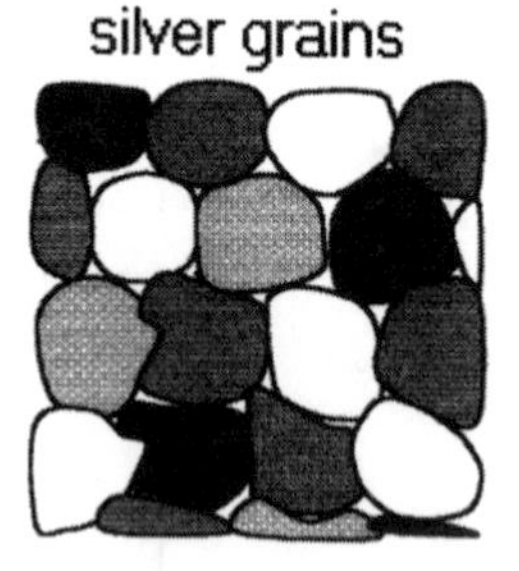

Resolution of a raster scan (left) and photograph (right).

As the denominator of the photo scale increases, the area on the image increases by quadratic increments (i.e., the area covered on a 1"=400' photo is four times the area covered on a 1"=200' photo). In this sense, four times as many ground objects are squeezed into each silver particle in the smaller scale photo. Because of this compression, the definition of each object decreases and finer details are lost. On a 1"=200' photo you may be able to identify objects as small as tar cracks in streets, hydrants, trash cans, street signs, mailboxes, and other features of similar magnitude. However, you will fail to identify such objects on a 1"=2,000' photo.

In addition, decreasing pixel size will not markedly improve resolution. If an object is too small to view at a given resolution, you cannot resample the pixels and fix the problem. For example, the smallest feature a silver grain can detect on a small-scale photograph may be the size of an automobile (at a pixel size of 10 ground feet). However, if your goal is to locate inlet grates (at a pixel size of 1 ground foot), you would be unable to resample the pixels to locate the grates, because the smallest feature the silver grain "sees" is still the size of an automobile.

To illustrate this point, refer to the next image. The sketch on the left shows objects detectable on an image at a specific pixel resolution. On the right, the same image area is scanned at a higher pixel resolution. Although the visual quality of the image may improve somewhat, you still see the same objects in relative accuracy with one another.

Example of changing pixel resolution within a specific scene.

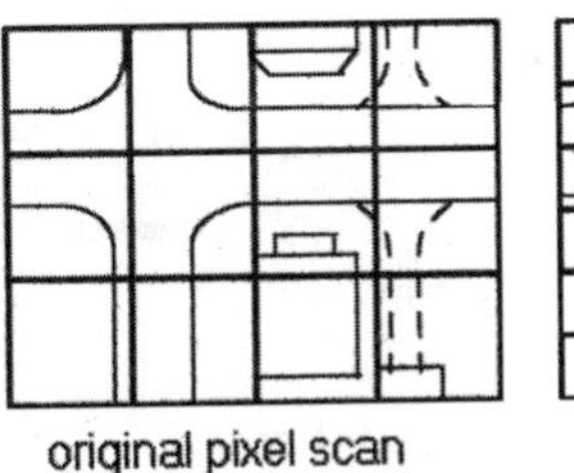

Camera Technology

Cameras used for aerial photography are mechanically and optically precise and therefore very costly. The following figure illustrates a cut-away section of the camera, showing the focal plane where the film rests during exposure, the compound lens system, and the shutter between the upper and lower lens elements. Focal length is the distance from the nodal point in the lens element to the focal plane. Although there are some special purpose cameras with different focal lengths, most mapping cameras have a fixed focal length of

6". A detachable magazine that holds the film roll is seated atop the camera cone during the photographic mission.

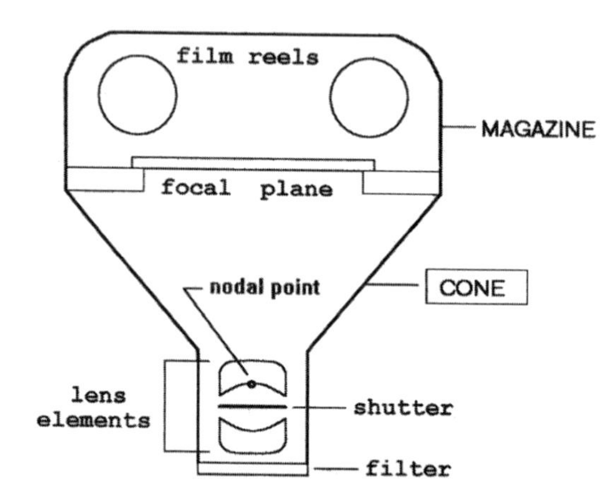

Cross section of an aerial camera.

A standard exposed image negative typically measures 9" square. The next figure illustrates the pair of congruent triangles (A and B) that form a fixed relationship between the area covered on the ground and the same area registered on the film negative. Triangle A represents the camera cone, while the base of triangle B is the mean project elevation.

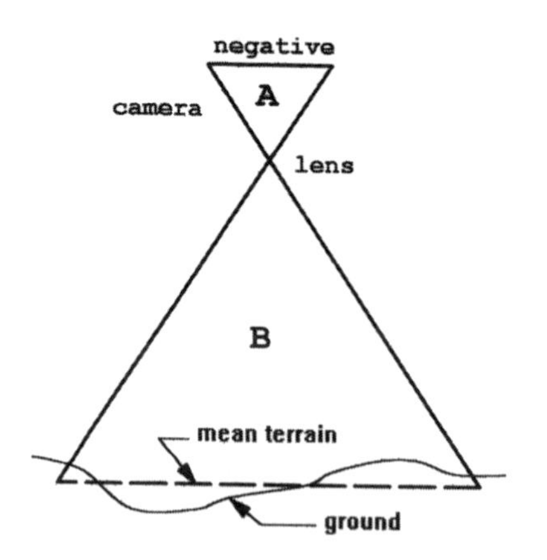

Congruent triangles involved in aerial photography.

Digital Frame Cameras

Recent technological developments in digital frame cameras (DFCs) and associated charged couple devices (CCDs) have made DFCs suitable for aerial photogrammetry. The following figure depicts a DFC.

Schematic of DFC system.

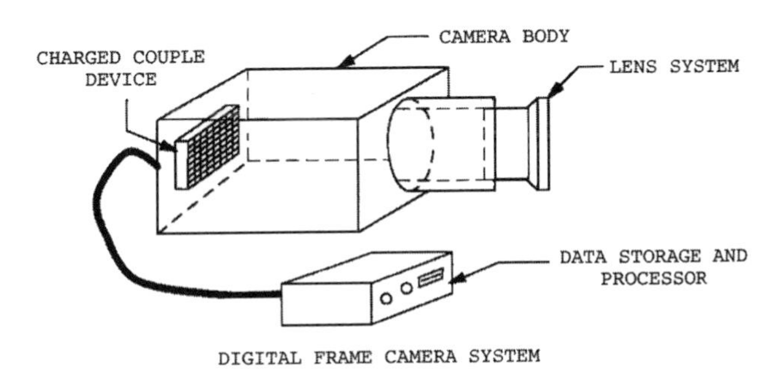

Reconnaissance photography and small area mapping are currently being accomplished with this type of system, which allows for high resolution data to be captured and processed relatively quickly. Until recently, DFCs were not calibrated and did not provide a sufficiently large CCD to capture enough ground area to be economical. However, increased CCD size and computer processing speeds now make DFCs competitive with conventional aerial film and videography. The resolution and area of coverage of a single frame increase with the size of the CCD. Because DFCs are relatively small, potentially less expensive, and virtually offer real time data availability, usage will likely increase in the near future as development continues.

Aerial Photography

Notes on Displacement/Distortion

It is a common misconception that aerial photos are identical to maps. Chapter 13, "Scanning," outlines methods for digitally manipulating a photo into a pictorial or linework map, but it is misleading to refer to an uncorrected aerial photo as a map.

It is important to keep in mind that an aerial photograph is a two-dimensional representation of a solid 3D scene. It is unwise to use a single aerial photo as a precise map due to inherent terrain displacement. Because a photo image is the result of rendering 3D features onto a two-dimensional plane, terrain relief shifts features on the photograph away from their true orthogonal position in a correctable pattern. The greater the change, the greater the image displacement. The figure below (left) depicts the amount of displacement (d) on photos of a tower taken at two consecutive camera stations, where the top and bottom of a tower have the same horizontal coordinate but a different elevation.

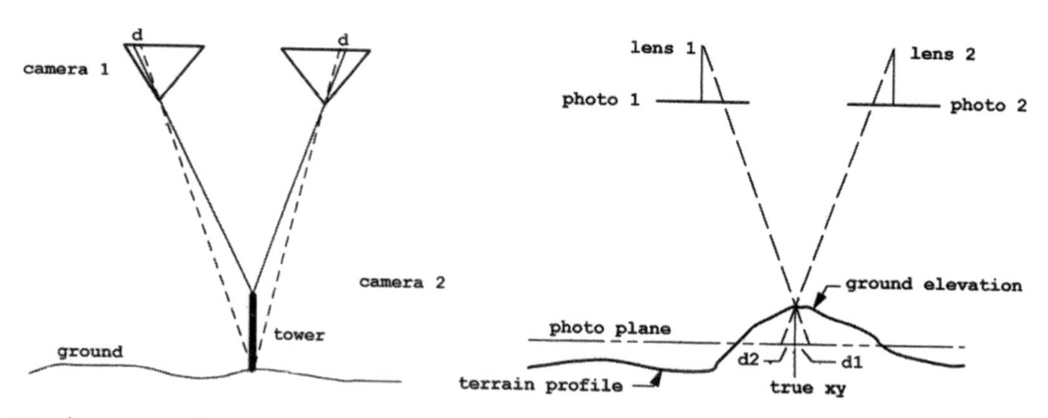

Displacement of an upstanding cultural feature on an aerial photograph (left). Displacement of a terrain feature on an aerial photograph (right).

Displacement shifts positions of terrain features on the photograph in the same manner as it does upstanding cultural features such as towers, chimneys, buildings, and poles. The illustration above (right) shows how a rise in the terrain displaces an image. In order for an image point to be in its true horizontal geographic, it must be at its true elevation.

Because displacement is inherent, it does not indicate a flaw in a given photograph. Moreover, displacement can be predictably corrected in the mapping process. Indeed, without this phenomenon, it would be impossible to create a map depicting three dimensions. When each eye is focused on a particular image feature, such as that seen by the camera from two different aspects (parallax effect), the mind interprets it as seeing a single image with three dimensions. Because photos of ground objects are not depicted in their true positions, you cannot use them as precise maps.

While displacement is a natural occurrence, distortion does indicate one or more malfunctions in the photographic or film processing systems. Precision mapping cannot be accomplished with distorted photos because rectification is unpredictable. However, distortion rarely occurs when modern photogrammetric techniques are properly applied.

Photogrammetric Techniques

Aerial photography requires that a precision camera be mounted in an aircraft with a hole cut out of the floor. This allows the camera to be mounted so that it looks directly down at the ground. These cameras are expensive and usually cost far more than the airplanes that carry them. A flight crew is routinely made up of two individuals, a pilot to fly the aircraft and a photographer to manipulate the camera. The figure below shows the seating arrangement for a two-person flight crew. On an airborne global positioning system (GPS) flight, the pilot and photographer share extra duties to control the GPS system. .

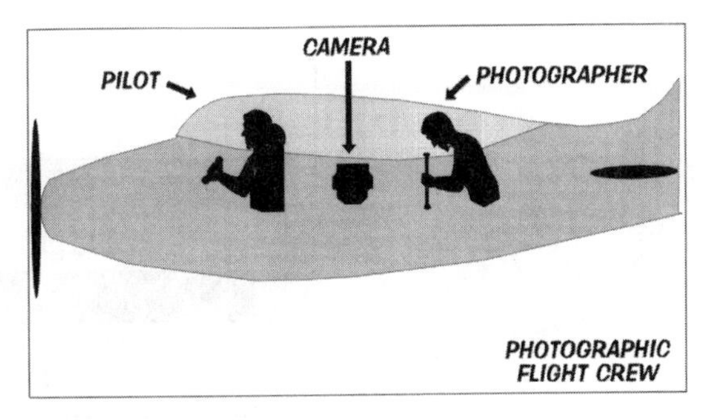

Seating arrangement of aircraft crew and camera in a photo flight. (RM)

In order to create maps that truly represent the Earth's surface, the 3D effect of overlapping aerial photographic images must be employed. To accomplish this, the location of each flight line is preplanned to ensure that no gaps in photo coverage exist between adjacent strips. This can be accomplished using manual or electronic methods. On a mission not controlled by GPS, the centerline of each flight line can be manually delineated on an existing map, and then the pilot navigates by visual reference to ground objects. Some aircraft are fitted with autopilot features linked to airborne GPS that rely on a computer to guide the flight path along a predetermined course.

Most mapping projects require that adjacent flight lines overlap one another by 30 percent of the photo width. The following figure shows the typical sidelap area between two adjacent flight lines.

Head-on view of sidelap between two adjacent flight lines.

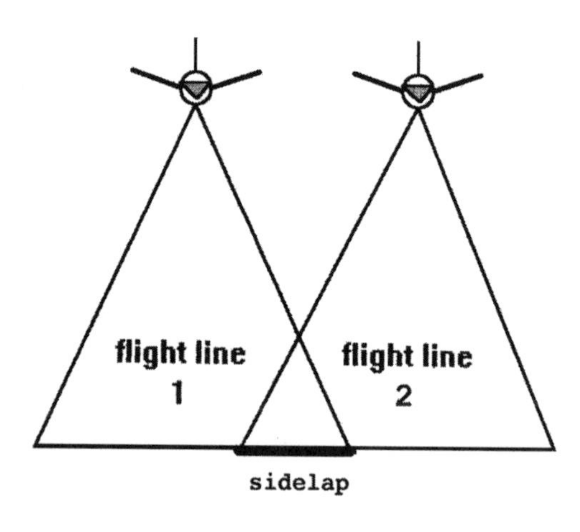

Sidelap areas involved in an aerial photo project.

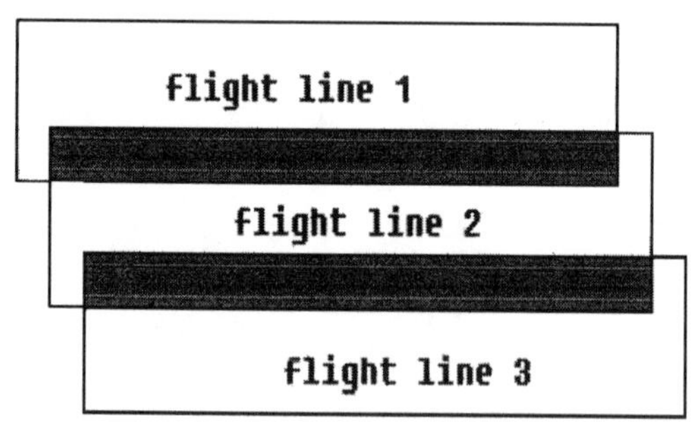

The previous figure illustrates the image coverage of a series of photo lines and the resultant sidelap between adjacent flight lines.

As the airplane flies along each designated line, an electronic system automatically activates the camera shutter at periodic intervals to ensure that sequential images with sufficient endlap are exposed. Most mapping projects require that successive photos in a flight line overlap one another by 60 percent of the photo width, although some orthophotos may demand a larger endlap, perhaps by as much as 80 percent. The figure below indicates the area of normal endlap between two successive photo images.

Endlap in a pair of successive photos in a flight line.

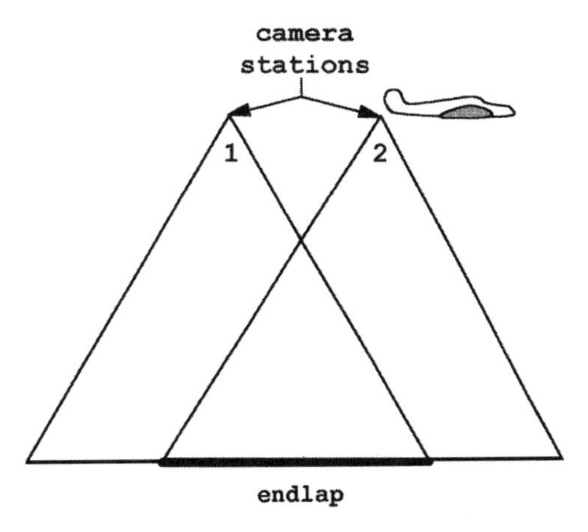

This common overlapping area of two successive photos, known as a *model* (or stereomodel or stereopair), produces the 3D effect that allows for the creation of maps. The following figure delineates the model area on two successive aerial photos.

Model area.

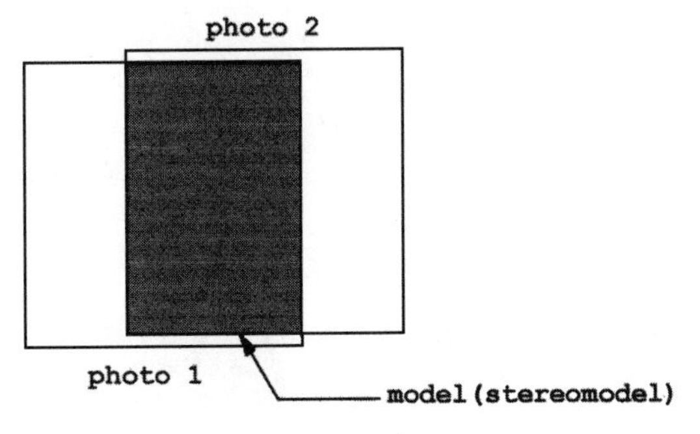

Assuming normal endlap and sidelap conditions, the next figure demonstrates the amount of 3D coverage that each sequential pair of photographs contributes to the total project. In technical parlance, the area shown is known as the *neat model*. In a multi-model project, this is the area the photogrammetric technician works with to create a map.

Neat model area of a stereopair.

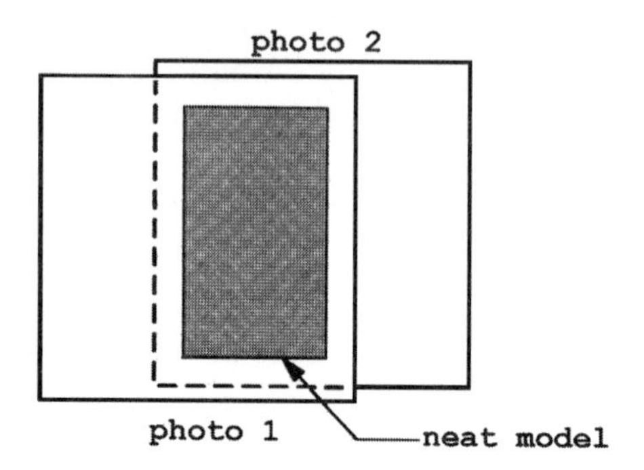

Flight Conditions

Conditions under which aerial photo missions are undertaken vary according to the purpose of the photography and the environment of the project area. Because flights usually require full sunlight, cloud-free days are preferred. Shadows can obstruct ground detail, so the time of flight should ensure relatively short shadows. This limits the period when photos can be taken to the three hours after sunrise and three hours before sunset on unclouded days. Ideally, aerial photo missions also require that the sun's angle be greater than 30 degrees above the horizon.

Because contour mapping requires that the stereo technician be able to see the surface of the ground, flying may be limited to the season between November and April when foliage is minimal. Snow cover and sun angle in the northern climates may restrict this period considerably.

As a rule, color film requires more favorable weather conditions than panchromatic (grayscale) film. On the other hand, infrared film is more proficient at penetrating haze.

As a safety precaution for the flight crew and general public, the minimum flight height is typically restricted to 1,000'. Maximum altitude of a mapping aircraft is limited by its mechanical configuration. Although an aircraft may be capable of flying above 18,000', Federal Aviation Administration (FAA) regulations are most strict regarding the airspace above 18,000' because of heavy airline traffic. Hence, most commercial aerial photography flights take place between these altitudes. In special situations, a helicopter may be used as the camera platform, and lower altitudes (and larger photo scales) may be attained.

Aerial Film

Several types of aerial film can be used for aerial photography, depending on a project's requirements. These films are sensitive to a narrow band of wavelengths.

All types of aerial film are composed of a layer of light sensitive silver salts spread over backing of stable plastic. Minute particles of silver react chemically to the amount of radiation they are exposed to, which forms an image. Panchromatic (black and white, or grayscale) film has a single emulsion layer that reacts to all colors of visible light and produces a grayscale image. The following figure shows a cross section of panchromatic film.

Cross section of panchromatic film.

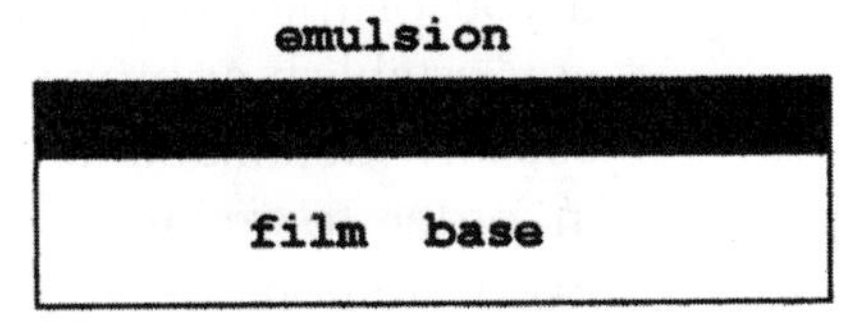

Infrared film has a single emulsion layer similar in structure to panchromatic film. It reacts to near infrared radiation (reflected heat) and produces a grayscale image.

Color film has three dye layers that react to blue, green, and red visible light, yielding a true color composite image. The next illustration shows a cross section of color film.

Cross section of color film.

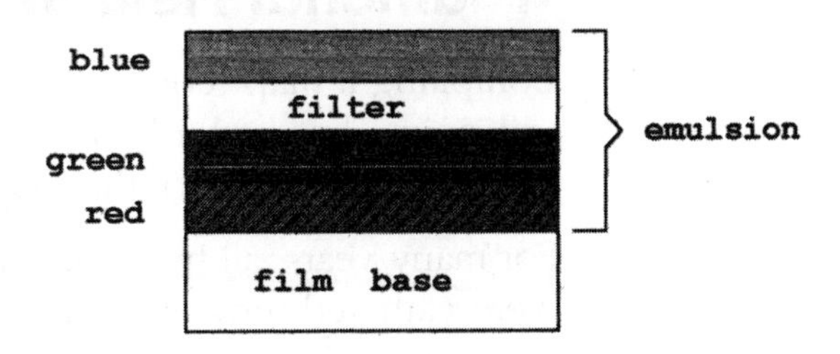

Color infrared (or false color) film has three emulsion layers that react to red or green visible colors or the near infrared portion of the electromagnetic spectrum and produces an image of hues that do not truly depict the colors on the ground. For instance, vigorous vegetation appears in shades of red, plowed fields of red clay appear as green, while water appears from light blue to black, depending on the amount of suspended particulate pollution.

Panchromatic and infrared films are developed to a negative, which can then be processed into several positive products. Natural color film can be processed into a negative or color transparency, both of which can be processed into positive products. False color film is usually processed to a color transparency and then into positive products. Once the film has been processed, it is titled at a minimum with the date of the photography, project identification, roll number, and photo number.

Photo Control

If you plan to map from aerial photos, it is imperative that you establish appropriate grid coordinates for several specifically identifiable terrain features visible on the imagery, in order to reference the photo to ground coordinates. Every stereomodel requires a minimum of four strategically placed photo control points to reference it to the terrain. A photo control point is a specific ground feature that can be located and identified on the photograph, and for which a ground coordinate has been established. Points may take the form of targets placed on the ground prior to the photo flight or discrete identifiable image features.

Traditional Field Surveys

Compiling a map requires that horizontal and vertical control points be established to accurately validate the scale, azimuth, and datum of the map.

For many years, all horizontal and vertical mapping controls were gathered through traditional ground surveys. The surveys required that traverses be run through a series of courses using theodolites to turn angles, graduated steel tapes or electronic range finders to measure distances, and spirit levels to calculate elevations. In recent years, the advent of the total station allowed these procedures to be more efficiently accomplished by a single instrument, although the procedural routines remain similar. These techniques determined the X (easting), Y (northing), and Z (elevation) coordinate triplets on control points. At present, the traditional field survey may be the control procedure for small projects.

The following figure indicates that a minimum of 32 field survey points would be required to control a mapping project consisting of 14 models on three flight lines by gathering all data on the ground.

Location of field survey points required for mapping if controlled by conventional field surveys.

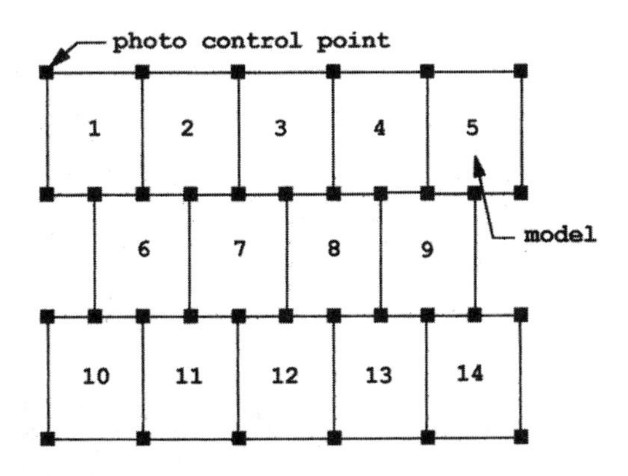

GPS Field Surveys and Aerotriangulation

Currently, employing space based GPS procedures to establish the coordinates of field survey control points is more common than using traditional survey methods. See Chapter 12, "Global Positioning Systems," for more information.

As the science of navigational and surveying technologies has advanced, the procedures necessary for defining geodetic position and surface elevation have been periodically refined. In a GIS query, you may access several map layers of a project that have been gathered from different sources. Although each may be accurate, the datum reference frames may be different, which can lead to difficulty matching geographic reference points. As a result, the layers would not overlay one another accurately and therefore yield a false solution. The metadata in the file layers should explain the data set's mapping datum.

Aerotriangulation is another common process for establishing control points. Also known as artrig, analytical control bridging, or AT, aerotriangulation is recognized for its financial and scheduling advantages. It also requires fewer actual field control points per project. Of course, there is a cost involved in providing the balance of control points necessary for your project, but it is less expensive per unit than field surveys. Another, perhaps equally important, advantage is access to ground points. Considerations such as terrain character, private property exclusion, and site inaccessibility can preclude the establishment of a ground point at a desired location. With airtrig, these roadblocks can be overcome.

In the previous illustration depicting field survey points, 32 points are required. The next figure indicates that 12 or fewer field survey points are required using aerotriangulation for the same project.

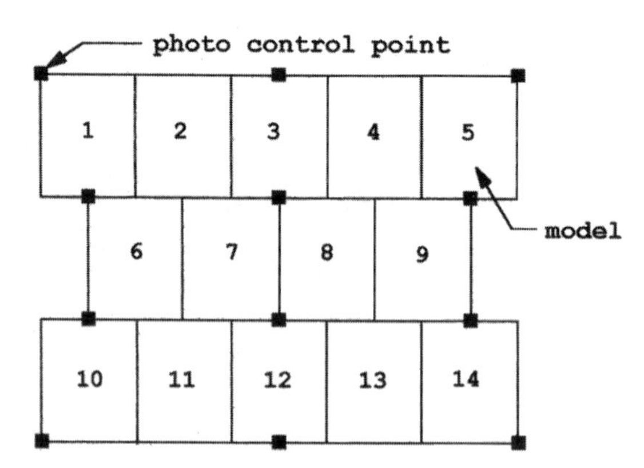

Location of field survey points required for mapping if controlled by aerotriangulation.

All supplementary control points required to orient the models are selected as synthetic photo image points. In brief, on each photo three points are selected. By transferring points from each surrounding photo, this system produces a minimum of six essential photo control points on each steropair. Through the use of a sophisticated mathematical package, terrain coordinates can be assigned to each of the selected image points. The number of photo control points required to orient each photo to the Earth's surface is shown in the following figure.

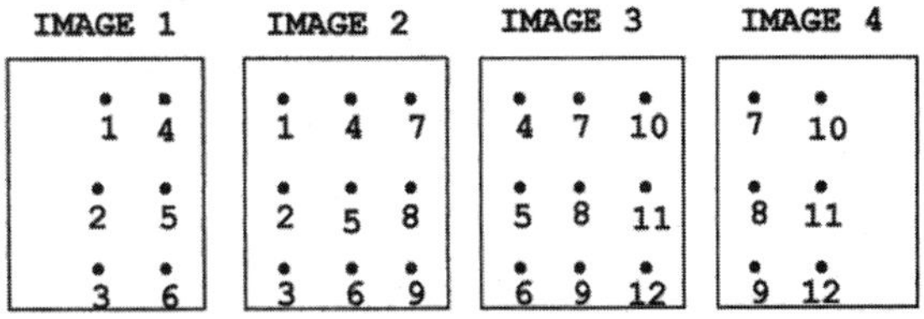

Pattern of photo control points required in aerotriangulation.

IMAGE 1	IMAGE 2	IMAGE 3	IMAGE 4
1 4	1 4 7	4 7 10	7 10
2 5	2 5 8	5 8 11	8 11
3 6	3 6 9	6 9 12	9 12

Airborne GPS

A fairly recent innovation in the science of photogrammetry is the merger of aerial photography, supplementary photo control, and GPS. This union allows instantaneous position reckoning with a real time kinematic technique. The next figure shows that the GPS receiver in the aircraft locks onto several satellites to determine the spatial position of the camera. At the same time, a ground receiver at the location of a known coordinate point is locked onto satellites, periodically computing its position. Because its true position is fixed, the variables in the periodic signal range during its observation tenure can be applied as corrections to airborne camera positions.

Airborne receivers continuously gather GPS information. Because spatial location is defined by precise time at every occurrence of shutter activity, an electronic pulse triggers the GPS receiver to record the time of exposure. This information can be input to the aerotriangulation process to furnish sufficient photo control for a mapping project. A few ground targets are also provided to correct the datum of the calculation coordinate to the specific terrain level of the project.

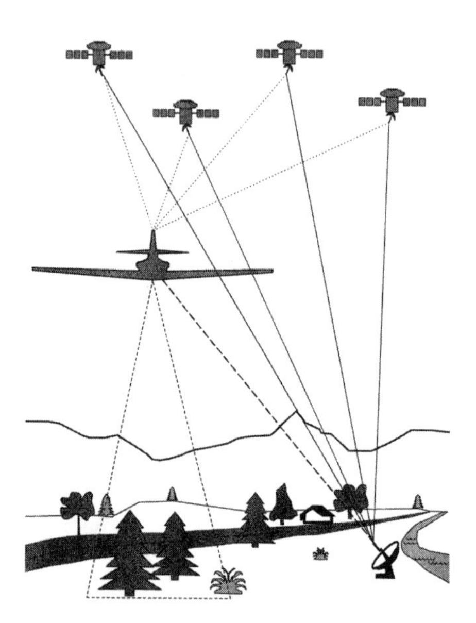

Locating the camera's spatial position with airborne GPS. (DF)

Just as aerotriangulation requires fewer terrain points than conventional ground control, airborne GPS further diminishes the number of necessary field survey points. Square or rectangular projects can be accomplished with as few as a single ground point at each corner of the photographed area. Linear or irregularly shaped projects demand more ground points. The location of ground points is also more flexible in this system.

Airborne GPS flights may require more planning than conventional flights. Typically, airborne GPS flights involve photography with 60 percent sidelap, sporadic cross flights, and lock-ons to at least five satellites. However, precautions vary with required photo and map scales and experience of the photogrammetrist.

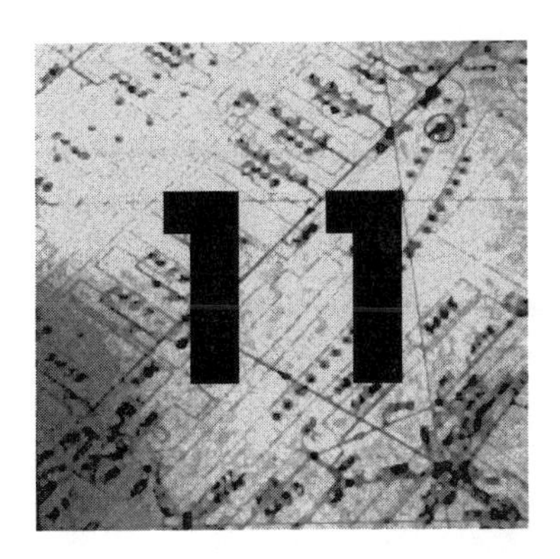

Producing GIS Data from Aerial Photos

Edgar Falkner and Dennis Morgan

Stereomapping

Prior to the first stereoplotters, mapping was performed on the ground using a plane table (a square board mounted on a tripod) that held a fixed sheet of paper. Ground features were plotted on the map sheet by sighting to the objects with an alidade. The surveyor moved to various vantage points and drew the map by hand. This procedure was extremely labor intensive. However, beginning with the Great Depression, map makers began to utilize aerial photos. Since then, stereoplotters have almost exclusively replaced surveyors. A stereoplotter is an instrument that accepts a pair of overlapping photos and in return, produces a spatial image that allows an operator to compile precise planimetric and topographic maps.

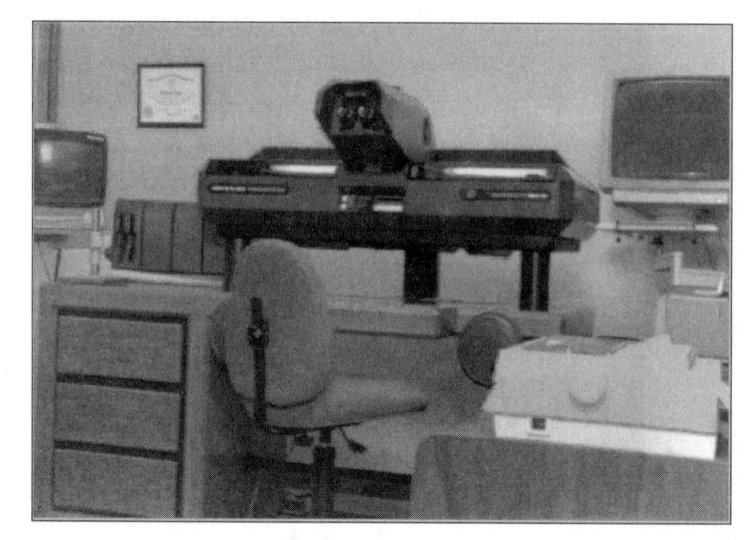

State-of-the-art, fully analytical stereoplotter system. (PS)

Planimetric map details are natural and manmade cultural features located according to horizontal positions (e.g., buildings, roads, railroads, bridges, overpasses, water features, treelines, utilities, and transmission structures). On large-scale engineering maps, features such as hydrants, inlet grates, parking meters, electric and telephone poles, mailboxes, manholes, and street signs would be included. Premarking of planimetric features prior to the flight is necessary when an object is difficult to identify on the aerial photo (e.g., a manhole, drainage outlet, or culvert). Sometimes these features are marked to identify the object and its type (e.g., "S" for storm sewers and "MH" for "manhole"). Premarking requires more initial coordination but can allow the photogrammetric technician to create a more complete map and eliminate field survey time and cost.

Topographic features indicate terrain relief with contours and spot elevations. Contours may be generated by either "dragging contours" or generating mass points and breaklines. Dragging contours is the process of stereoscopic viewing of the stereomodel and digitizing isolines of constant elevation. Collecting mass spatial points can be conducted in random or fixed grid order. When contours are computer generated with mass points, they are smoothed between adjacent points. In order to maintain the terrain's true character, breakline data must be collected, which involves selecting a string of points along a feature (stream bottom, ridgeline, or edge of road). When mass spatial point and breakline files are composed, contours can be generated with available algorithms.

Stereomapping machines have undergone at least three previous major generations: analog, analytical, and digital. The first machines were completely manually operated, and map preparation owed as much to the operator's artistic talent as to the science of photogrammetry. More recently, compilation work is shared by people and machines. The figure below represents how a stereoplotting system generally functions in the process of collecting spatial vector mapping data.

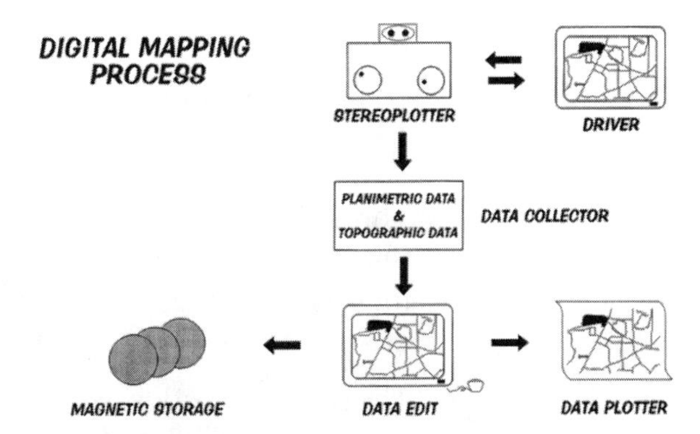

General procedure for collecting spatial vector mapping data with a stereoplotter. (RM)

- In order to create mapping information, two successive overlapping photos are placed in receptacles inside the stereoplotter. Each photo is manipulated so that it assumes the same position it had when the camera took the photo. With a coordinated process between the technician and the driver computer, the photos are brought into proper alignment to create the 3D spatial image. The computer then georeferences each photo control point appearing on the model to the ground datum in a mathematical configuration. Once this is complete, every point on the stereomodel is in its true spatial location.

- Spatial data are extracted from the photos manually or via computer. The amount of effort required by the technician depends on stereoplotter technology. The data are in the form of individual spatial points, each consisting of a unique identifying feature code or number and a geographic location. The data are stored in a data collector, which may be internal or external to the machine. At some point, the information is subjected to an edit to check validity and correct erroneous data. Final data can be output in hardcopy map form and/or to a suitable storage media.

Topographic Detail

The following four phases are combined to extract topographic information from the stereopair and translate it into contours.

1. Employing a process known as autocorrelation, the computer within a stereoplotter creates a grid base whose dimensions are specified by the mapping technician based on the character and gradient of terrain relief. Through a sophisticated protocol, the computer determines the ground XYZ coordinates at the same grid intersection on each photograph. This results in a digital elevation model (DEM) file.
2. The operator adds supplementary data strings along irregular breaklines (e.g., stream bottoms, ditch lines, and ridgelines) and spot elevations on hilltops, saddles, and road intersections. This collection of terrain breaklines (strings of individual XYZ points) is known as a digital terrain model (DTM).
3. Once terrain data have been accumulated, complex mathematical software can form a triangulated irregular network (TIN) of data points to create a base file of interpolated points on specific contour interval values.
4. By inserting the TIN file into a contour software package, the computer can connect points of equal elevation

and generate contours. Inexpensive DEM files are available from the U.S. Geological Survey (USGS), but they are appropriate for large contour intervals only (10 or 20' intervals and larger).

The files discussed here are created to generate topography accurate to a specific contour interval and vertical accuracy, but the computer can take this information and generate contours to any interval. However, just because the computer has this function does not mean the results will be accurate.

Consider another example of accuracy standards for vertical data. This time, assume you have a DTM file created to generate contours at 10' intervals. Depending upon which accuracy standard is applicable (see the "Accuracy" section later in this chapter), the generated contours should be accurate from 3.3' to 10'. The computer can readily accept these base data and generate contours at 1' intervals. However, you cannot assume that the vertical accuracy tolerance will be within the 0.3 to 1' range required under the mapping standards for 1' contours. Rather, the expected error of these contours would still be in the 3.3 to 10' range and beyond the accuracy limits allowed, because the DTM file was created to generate 10' contour intervals.

Planimetric Detail

Digital data for planimetric (or cultural) detail are collected in vector format. (see Chapter 2, "Data, the Foundation of GIS," for more information.) Collecting data for cultural features is accomplished by the operator physically guiding a reference marker, an intrinsic feature within the machine that indicates the spatial position of a point—while viewing the 3D image. A segment of data is identified by a signature that the computer recognizes as a symbol for a specific feature. The computer then draws the symbol on a hardcopy map sheet. For example, digital data for a simple building would include a feature code plus four individual data points. When drawing the symbol, the data printer would draw straight lines connecting the four points to form the footprint of a rectangular structure.

This process results in the graphic output of a hardcopy planimetric map. Many projects are currently substituting rectified images or orthophotos for symbolized maps. By merging this scanned raster file with a DEM vector file, a base of XYZ coordinate points is created to generate an orthophoto image as accurate as a map. For more information on scanning, see Chapter 13.

Inexpensive digital line graph (DLG) files are available for purchase from the USGS. These data include hydrologic, transportation, and political boundary features. Keep in mind that these files are accurate only for mapping scales 1"=2,000' or smaller. It is also wise to remember that DLG files are created to produce cultural detail to a *specific horizontal accuracy*, but the computer can accept any DLG file and reproduce it at any scale. For example, you cannot assume that a DLG file created to generate detail for a map to scale 1" = 1,000' will adhere to accuracy standards if used in a map to scale 1" = 100'.

Softcopy Mapping

"Softcopy" is the term often used to describe the cutting edge of stereomapping technology. Softcopy instrumentation is a versatile union of stereoplotter and CAD technology. Now in its fourth generation, softcopy hardware and software are integrated into a CAD/CAM engine. At a basic level softcopy instrumentation is a hybrid mapping system combining a high processing speed workstation with stereoplotter components. The next illustration shows an image of a typical state-of-the-art softcopy mapping system.

Softcopy mapping system. (PS)

Tip and tilt factors at individual camera stations (calculated during the aerotriangulation process) and lens distortions (noted on the camera calibration report) are entered into the orientation file along with the airtrig point coordinates. The scanned raster files from two sequential images are displayed as a georeferenced stereopair on the monitor screen in an apparent 3D mode. The technician then roves the stereomodel and captures orthogonally corrected data. Through a complex set of restitution algorithms, the central processor constantly corrects for displacement and camera aberrations on the fly as the reference marker moves around the image.

Uses

Raster and vector data can both be used in this sophisticated process, and the computers that drive softcopy plotters are capable of extracting substantial amounts of digital data from photos. This type of instrumentation is capable of but not limited to the functions listed below.

- Digitizing vector information from maps, plats, drawings, and monocular and stereo images.
- Accomplishing aerotriangulation procedures.
- Extracting thematic polygons from existing maps, photo images, or satellite screens, and creating or recreating thematic maps.

- Generating DEM files.
- Creating contours from imported DTM files or internally created files.
- Producing wireframes from files containing a series of digital XYZ vector points.
- Generating orthophoto images.
- Draping scanned raster image data from a variety of remote sensors over wireframes to create perspective scenes.
- Creating aspect slopes from imported DTM files or files collected internally.
- Propagating line of sight profiles.
- Acting as a raster/vector link that can reference various database layers to solve GIS queries.

Increased interest in this type of mapping has generated many techniques and related mapping products that were unavailable a short time ago. Once the image is stored in a computer capable of softcopy mapping, then digital orthophoto mosaics, screen digitizing of planimetric features, and image draping are relatively easy to accomplish.

Hardcopy Output

Softcopy mapping plans usually require some in-house hardcopy output. One of the reasons for the increased use of softcopy mapping systems is the ability to produce economical hardcopy plots as necessary in a short period of time. Areas of interest can be enlarged in the softcopy mapping system, and then thematic maps can be generated and sheeted, producing hardcopy plots. In addition, producing quality paper plots of orthophotos on plotters is generally cost effective because the cost of computer processing time has declined and processing methods have become more efficient. Furthermore, be aware that lower quality plots have limited use and resolution quality can suffer. High quality plots require a piece of hardware and associated software known as a film writer. A film writer processes a digital orthophoto image and produces a film negative. The film negative can then be taken into a photo laboratory and enlarged to a specified horizontal map scale.

Image Analysis

A companion discipline to photogrammetry, image analysis combines the engineering aspect of stereomapping (to create a precise map base) with the analysis of image features (to create and identify thematic polygons). Previously, this process was limited to air photo interpretation, but currently the image may also be in the form of one or more segments of the electromagnetic spectrum, such as the total range of visible light, near infrared segment, several bands of thermal, or several bands of radar. Photomapping technicians employ a form of image analysis when identifying and compiling planimetric features. Image analysis is called into play when the user has a need for a polygonal map delineating pertinent themes, which can include—but are not limited to—soils, agriculture, geology, hydrology, forests, land use, utilities, transportation, urbanization, school districts, municipal boundaries, recreation, or transmission facilities.

Some projects may be limited to visual feature recognition by qualified technicians. Others can be adapted for computer recognition. Data scanning assigns each picture element a radiometric reflectance value ranging from zero to 255 (see the next figure). Black is zero brightness (no reflectance) and white is 255 (full reflectance), with 254 gradations of grayscale (partial reflectance) in between.

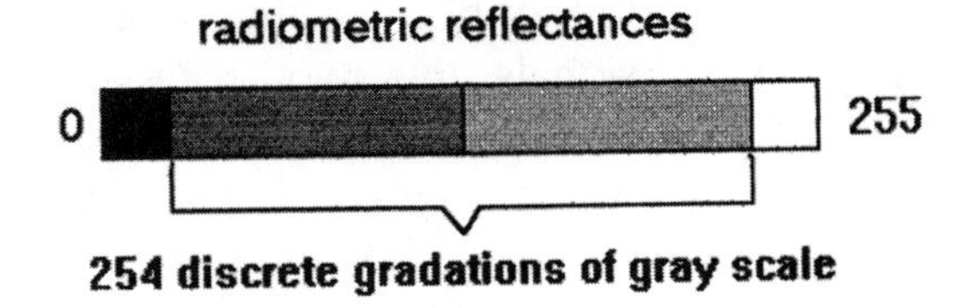

Range of radiometric densities.

In an unsupervised mode of recognition, the computer breaks the radiometric range into the number of segments specified by the user, then selects all features exhibiting similar discriminate radiometric signatures. A major disadvantage of this process is that the computer does not identify the polygons it designates, so the user must classify them through expensive field investigation or another method.

A more pragmatic approach is the supervised mode. In this process, a technician prompts the computer with samples of known target signatures, and the computer searches for similar radiometric reflectances, producing a map of designated thematic polygons. This procedure is useful because the user already knows what each polygon represents in the land use scheme—a factor that may substantially reduce the time devoted to expensive ground truth verification.

Because the Earth's landscape is dynamic, an important offshoot of electronic image analysis is the process of detecting change by revisiting data of a specific scene. This procedure allows you to scan an image and construct a base of unique thematic polygons, then repeat the process with another time lapse image of the same area. After they have been georeferenced to the same geodetic base, the computer compares the radiometric value of each pixel in both data sets and flags those that indicate a change in value. The first image may depict a pasture on a specific tract of land, while the second shows a block of newly built residences. Noting the differences, the computer produces a polygon of the affected area.

Digital Orthophotography

Orthophotography is often misunderstood among professionals and non-professionals in GIS and mapping fields. To help define the term, it may be helpful to learn what an orthophoto is not. It is not a scaled photo enlargement that exhibits unrectified image displacement. Nor is it an image that has been warped or "rubbersheeted" to an existing map source or coordinate system that also contains unrectified image displacement. A simple definition of an orthophotograph is a pictorial image processed to correct scale variations and produce an orthogonal image.

Orthographic Projections

The orthophotographic process has been commercially available since the mid-1980s. A major benefit of orthophotography is its ability to depict the relative ruggedness of an area in addition to basic elevation. An orthophoto image overlaid with contours makes this type of analysis possible.

In aerial photography, ground objects within a specific frame or snapshot are projected through the aperture of the camera and then onto film. This causes objects closer to the camera to be displaced radially away from the center of the photograph and to appear larger than similar objects that are farther away (see the next illustration). An orthophoto process corrects these displacements and produces an image map with consistent scale throughout.

Orthographic and perspective projections.

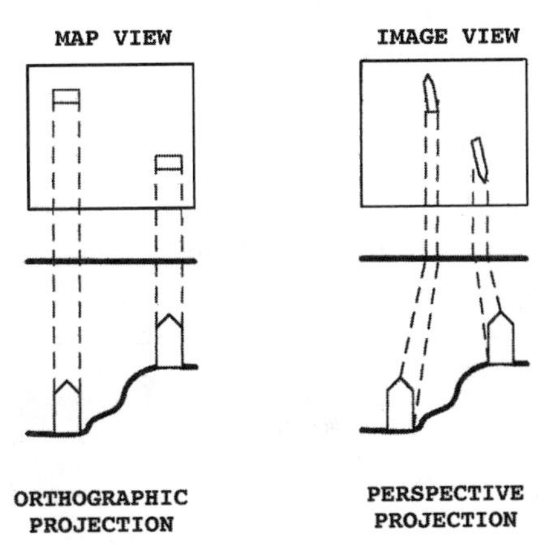

Digital Orthophoto Process

Orthophotographs were initially produced using analog methods. As stereoplotters became fully analytical, computer processing speeds increased, and the cost of hardware decreased, the development of high resolution scanners accelerated dramatically. These developments, from the late 1980s and through the early 1990s, hastened the development and use of digital orthophotography. Using this type of mapping is also on the rise because of relatively low equipment costs. The ability to view images at various scales, perform accurate image classification and change analysis, and develop high quality and low cost thematic maps for planning, presentation, and engineering design have made digital orthophotography one of the most popular data sources in GIS today.

The process of creating a digital orthophoto involves the use of traditional analytical photogrammetry and high resolution scanning. Although it takes less time (often significantly so) to generate a digital orthophoto than to generate a planimetric, topographic 3D vector map, orthophotography does require special, higher cost equipment and experienced personnel. High quality digital orthophotos require high quality aerial photography, an accurate elevation model, high resolution scanning, and the knowledge and experience of skilled aerial pilots, lab technicians, and photogrammetrists.

The figure below depicts the process in which vector data collected by a steroplotter can be merged with raster scanned images to create orthophotos with superimposed line mapping data.

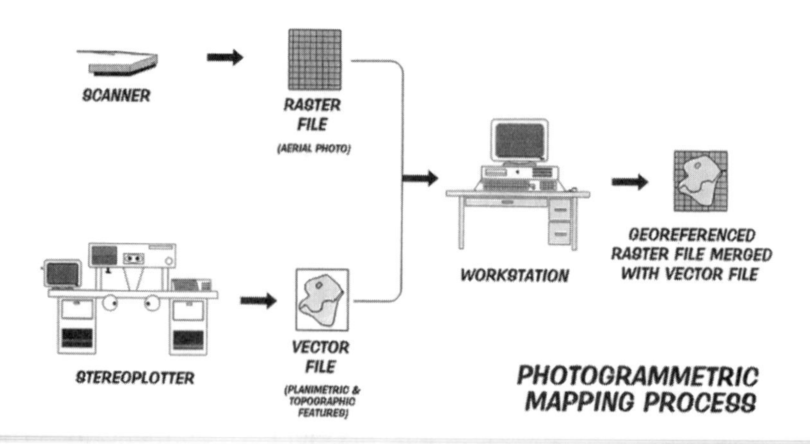

Raster/vector photogrammetric mapping process. (RM)

Planning the Flight

The first part of the process is planning the aerial flight, which includes considerations typical of every aerial mapping project: altitude, camera and lens type, endlap and sidelap, weather, and map accuracy. However, one critical detail that is often overlooked—and even misunderstood—is the ground resolution to be achieved in the orthophoto. Remember that you cannot improve the resolution of the original image; computer technology only allows you to manipulate and enhance the original image. Therefore, you must determine before the flight the resolution that your GIS will require.

For example, imagine comparing two sets of imagery for the same area, flown at different photo scales. The amount of land covered within a single pixel differs for each set. The larger scale orthophoto, flown at a lower altitude and showing larger objects, will show more detail within a single pixel than the smaller scale orthophoto. Depicting an automobile in the larger scale image may require several pixels, but it may require less than one pixel in the smaller scale orthophoto. You must understand this basic principle because it will affect the cost of acquiring the orthophoto, the computer equipment required to view and manipulate it in a GIS, and the accuracy of queries you can perform.

Once you determine the appropriate orthophoto ground pixel resolution for the aerial flight, you must determine the ground control and elevation model requirements necessary to meet a specified horizontal map accuracy standard. Remember that the map accuracy you specify affects the cost of your acquisition.

Field Work

Field work begins with placing ground control panels and acquiring the aerial photography. The standards you use should specify not only the altitude, overlaps, film and camera type, intended map scale, and other parameters, but also acceptable cloud cover flight deviations (i.e:, tilt, crab, and drift) in each image and flight line.

Translating the Image

After the imagery has been captured, it must be scanned and translated to digital format. Typically, this involves processing the negative to a film transparency known as a diapositive and scanning it.

Aerotriangulation

Scanned image files are loaded into a softcopy mapping system and aerotriangulation procedures are performed to establish the photo's position with respect to the ground.

Creating DEMs

When aerotriangulation is complete, DEMs can be developed. DEM data must be developed at a density compatible with the scanned image, a determination that requires expertise regarding the imagery, the equipment used to generate the data, and the intended use and expected accuracy of the orthophoto. Consequently, it is wise to consult an experienced photogrammetric firm for assistance. The anticipated horizontal accuracy expected from the orthophoto should be incorporated in design and development.

The last part of the orthophoto process is rectifying the image, aerotriangulation results, and DEM. Computer software is used to calculate the brightness value of image pixels and rectify each pixel to a DEM location. When this process is complete, the new image file will be accurately geographically referenced. Software can also be used to adjust contrast between two images. This technique is often used when more than one image is required to map an area and a digital orthophoto mosaic is produced.

Digital Orthophoto Mosaics

Digital orthophoto mosaics are common end products in orthophotography because many mapping projects will cover more area than a single stereopair. With proper hardware and software overlapping, orthorectified images can be merged, and the files can be cut into appropriate areas or shapes. Virtually seamless images can be generated from overlapping images if orthophotos are properly designed. However, keep in mind that achieving a seamless image demands using proper hardware, software, and design throughout the entire process (i.e., aerial photography, ground control, and orthophoto generation). Typically, seams result from poorly acquired imagery, insufficient and/or poor ground control design, low end software and hardware, and insufficient experience in DEM generation and orthophoto software usage.

Orthophotos as Base Maps

A properly designed digital orthophoto can serve as an accurate base map to collect selected digital planimetric detail with the use of screen digitizing. Once the orthophoto has been generated, digital mapping software can be used to create vector overlay files. The most accurate method of vectorizing planimetric features is to use mapping software that allows the user to view the orthophoto on a computer screen and use a cursor to digitize (or trace) the shape of features appearing on the screen. This is commonly called screen (or heads-up) digitizing. Although this method of data collection is often used to save money, remember that hardware and software affect accuracy. In most cases, cultural data collected in this manner are not as accurate as planimetric data collected using analytical or softcopy stereoplotters. However, in many cases this type of data collection may suffice. In either case, be certain to attach metadata files to the data.

3D Images

Three-dimensional images can provide planners and design personnel the opportunity to view a site from different perspectives. An orthophoto image and its digital elevation data can be processed and merged into a 3D model of an area in a process commonly known as image draping. With the use of special imaging software, a 3D model (or wireframe) is generated from the digital elevation data. The orthophoto image is then rectified to the wireframe. Software allows the user to view the model from any perspective; reshape areas; design, cut, and fill for construction; and view the results of future potential development.

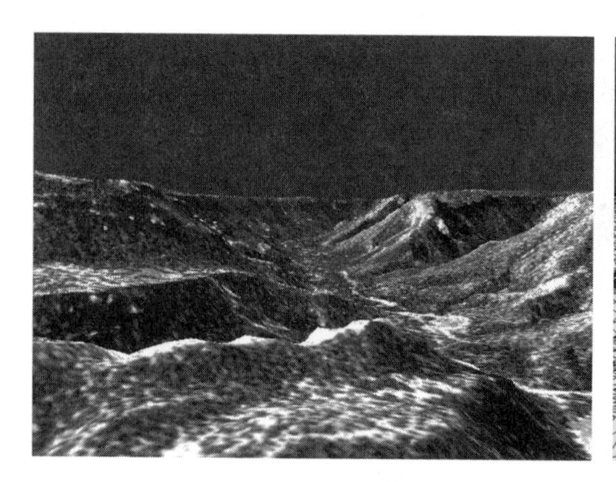

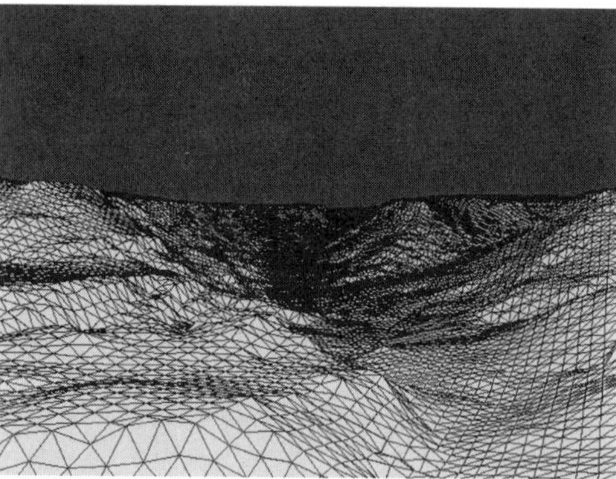

Draped orthophoto image (left) and wireframe. (PS)

Design Issues

When designed properly and used as intended, orthophotography is relatively problem free. An orthophoto should not exhibit displacement, distortion, or DTM (or DEM) aberrations. These problems occur when images are designed with techniques improper for orthophoto production. The quality of the DTM dramatically affects the map quality in terms of displacement, distortion, and other accuracy issues. The accuracy of the DTM is in turn greatly affected by the basic design parameters of an orthophoto project, and to a lesser degree, a map maker's experience. You must understand how each of these issues affects the overall quality of the final map. Producing aerial photography at the wrong altitude, inadequate digital elevation data, and poor scanning are common pitfalls resulting from funding constraints. Inadequate experience in map design, digital terrain and digital elevation data collection, and computer hardware and software requirements are common technical problems. Funding and technical problems can be avoided if you become educated in all aspects of the technology or purchase the services of a professional mapping firm. Issues regarding system throughput, film choice, and resolution are addressed in Chapter 13, "Scanning."

Project Planning Considerations

Minimizing the cost of airborne system data requires detailed planning. Project strategy should not only involve flight planning for data acquisition, determining appropriate spectral ranges, and agreement on processing techniques, but also searching for a qualified platform source. Developing a platform and sensor system specifically for your project is likely to be impractical. However, as demand for this type of service increases, choices among existing systems are rapidly expanding. You must take care to verify that the systems you are considering will indeed produce reliable data in an acceptable format for your project. The technical personnel should be familiar with sensor equipment and understand the technical aspects of your data. Reliable and productive systems allow sensors to be adjusted and transferred to other platforms rather easily according to project requirements.

Although photogrammetric mapping has advanced from an "art" form six decades ago to the "scientific" enterprise it is today, subjective project planning decisions can be made based on economics, timing, or convenience. In every situation, it is imperative that the map producer and user view product reliability as a goal. Both parties have a responsibility for ensuring that a map is accurate.

The surveyor or mapper has the right to expect fair compensation for this effort. By the same token, the user should receive a precision product for a reasonable price. Map producers and end users tend to approach project cost subjectively. Surveyors and mappers tend to include unpredictable "what if" scenarios in cost estimates for protection against financial loss, while end users can base their estimates on the amount of funds available rather than the effort necessary to furnish a quality product. Both parties must be realistic in pricing a given product.

Coordinate Systems

When mapping from aerial photos you must reference the imagery to a pertinent terrain coordinate base. The most common geodetic grid systems used in the United States are latitude and longitude, Universal Transverse Mercator (UTM), and state plane. Translation algorithms are available to convert one type of coordinate system to another.

Although they are not coordinate systems, three land subdivision strategies are used in United States to delimit land lines: metes and bounds, rectangular surveys, and Spanish land grants. Metes and bounds are irregularly shaped property boundaries used in the 13 original colonies, and the system of rectangular surveys system is used throughout the other states. The original Spanish land grants were plots of ground (usually a relatively narrow river frontage extending for a long distance inland) which the Spanish crown deeded to influential early colonizers. Within a system of standard parallels and guide meridians, all land is divided into named townships enclosing 36 sq mi each. Township and range lines segment land further into sections, each covering 1 sq mi (640 acres) and are numbered in a distinct sequence. Land within each section can be repeatedly halved or quartered into smaller parcels. The system of rectangular surveys is more common than metes and bounds or Spanish land grants.

Accuracy

Mapping Standards

National Map Accuracy Standards

From the early 1940s and through the late 1980s, the photogrammetric industry for mapping was dominated by federal mapping guidelines known as the National Map Accuracy Standards (NMAS). These standards recommend map accuracy for planimetric features and contours.

Simply stated, the NMAS maximum allowable horizontal error for 90 percent of cultural features that are field tested is 0.0333" at the intended map scale. For example, if you are testing the integrity of a map to scale 1"=200', 90 percent of tested planimetric errors must fall within 6.67 ground feet. Maximum allowable NMAS vertical error for 90 percent of field tested contours is one-half of the intended contour interval. For example, if you are testing the integrity of mapping with 2' contours, 90 percent of tested contour errors must fall within 1'.

American Society for Photogrammetry & Remote Sensing Guidelines

While NMAS are still used extensively today, the American Society for Photogrammetry & Remote Sensing (ASPRS) has published a set of guidelines for large-scale mapping that dictates error limitations for three classes of mapping.

The maximum ASPRS horizontal error for Class I field tested cultural features is an RMSE equal to the intended map scale denominator divided by 100. The error rate doubles and triples for Class II and Class III features, respectively. For example, if you are testing the horizontal integrity of a map to scale 1"=200', the maximum allowable RMSE for all tested objects would be 2 ground feet for Class I, 4 ground feet for Class II, and 6 ground feet for Class III.

Maximum ASPRS vertical error for Class I field tested contours is an RMSE one-third the amount of the intended contour interval. Again, the error rate doubles and triples for Class II and Class III features, respectively. For example, if you are testing the vertical integrity of a map with 2' contours, the maximum allowable RMSE for all tested contours would be .66' for Class I, 1.32' for Class II, and 1.98' for Class III.

Be aware that once digital data have been collected, scale and contour interval can be artificially manipulated by computer. In this context, the term "intended" is defined as the scale or contour interval anticipated to meet the specified standards. Although NMAS and ASPRS standards view the degree of allowable error from somewhat different perspectives, they both stress the user's responsibility for meeting certain accuracy standards for data layers.

Absolute and Relative Accuracy

Accuracy can have absolute and relative meanings. Absolute accuracy, or the accuracy of measurements within a given data layer, is a function of the smallest definable feature within the layer and the measurement capabilities of the equipment used to develop the layer. For example, absolute accuracy of an orthophoto is a function of pixel density; orthophoto measurements of area and distance can be no more accurate than the pixel size.

Relative accuracy is defined as the horizontal and/or vertical location of map features relative to a specified Earth datum. Relative accuracy for most topographic and planimetric feature data refers to the accuracy of procedures used to tie the data to the Earth (i.e., ground surveys). Ground survey accuracy requirements are generally specified in standards such as the Federal Geodetic Control Subcommittee (FGCS) Standards and Specifications for Geodetic Control Networks. Data sets may be accurate according to a set of standard specifications and/or relative to other similar data sets.

Most hardcopy and digital mapping data produced since World War II will state an accuracy standard or include metadata files. Vector mapping data, including planimetric features and topography, will state the standard used and horizontal and vertical scales when the map was compiled (e.g., NMAS for 1"=2000' with 10' contours). Raster map data may state a ground pixel resolution and a horizontal accuracy (e.g., 1"=100' with a ground pixel resolution of 1').

Because accuracy is a critical component of GIS data conversion, ensure that the metadata (or data about data) are as complete as possible for each data set. Metadata typically include the source of data, how and when they were compiled, intended uses, and their accuracy. See Chapter 15, "Spatial Data Transfer Standards," for more information.

Minimum Requirements

Data sources without metadata are questionable and should be used with great caution. The accuracy of your database is only as accurate as the least accurate data set. Minimum metadata requirements depend on the characteristics of the type of data. For example, tabular data such as census information should provide—at a minimum—source, minimum sample set, intended uses, and expected accuracy. Mapping data should provide the standard used, compilation methods and dates, intended uses, and the accuracy of the data at the original map compilation scale and/or contour interval.

Data accuracy implies a certain level of quality. The higher the accuracy the more reliable the data set will be—and, in general, the more time-consuming and expensive it will be because data cost is directly related to relative and absolute accuracy. Tabular data may require larger data sets at the outset for sampling and therefore require more processing time. Relative and absolute accuracy improvements in aerial photo mapping data require lower altitude photos, as well as additional ground surveys, compilation efforts, and funding.

As suggested above, funding is necessary to improve relative and absolute accuracy. Consequently, you must give accuracy requirements significant thought when planning your data conversion effort. GIS has a vital link to geography. All data sets are linked to geographic models, and metadata for the base geographic model should be as thorough and accurate as funding permits. Base data accuracy is critical because all other data sets will be geographically linked to the base model.

Source Data Manipulation

Avoid compromising source accuracy. With the advent of digital mapping data, and the relatively low cost of mapping softwares, it is becoming easier to convert mapping data intended for planning (i.e., 1"=1,000' horizontal scale with 10' contour intervals) into engineering design data (1"=50' with 1' contour intervals). Such conversions are not only illegal and unethical, they are ultimately useless because the accuracy level has not changed. You can resample a scanned image of an aerial photograph acquired for a pixel resolution of 5' and generate an image with a pixel resolution of 1'. However, the quality of the image will not improve. If your project requires that you resample a data set in order to improve compilation speed, the new data set should include a new metadata file explaining resampling techniques, relevant accuracies, and intended uses for the resampled data.

Global Positioning Systems

Richard Lewis

In the 1960s, there was no method for quickly and accurately locating positions of U.S. nuclear submarines. While information regarding target coordinates, ballistics, and missile trajectory was accurate, precise submarine coordinates were unavailable. In 1970, Congress authorized the U.S. Department of Defense to develop the NAVSTAR global positioning system (GPS) for $13 billion. The result is an orbiting satellite network capable of transmitting position information to the country's nuclear submarine fleet worldwide. The DOD degrades this GPS signal for non-military use through a process known as selective availability, providing civilians with 100m accuracy (2RMS, or 2 root mean square). There is a complementary process—differential correction (DGPS)—that restores some accuracy. However, commercial GIS users are among those civilians for whom the DOD has authorized accuracy of 5 to 15m or better from the GPS signal without differential correction.

It may seem complicated to locate geographic positions using a series of 24 satellites that orbit the Earth at an altitude of 12,600 mi and "rise" and "set" every 12 hours (see the next figure). However, GPS works in a method similar to trilateration (the measurement of distance and bearing), which is used to pinpoint a location for navigation purposes. If you want to find your location with a compass and map, you record four bearings to landmarks on the map. The result, or cross bearing, is your location. GPS works in a similar fashion.

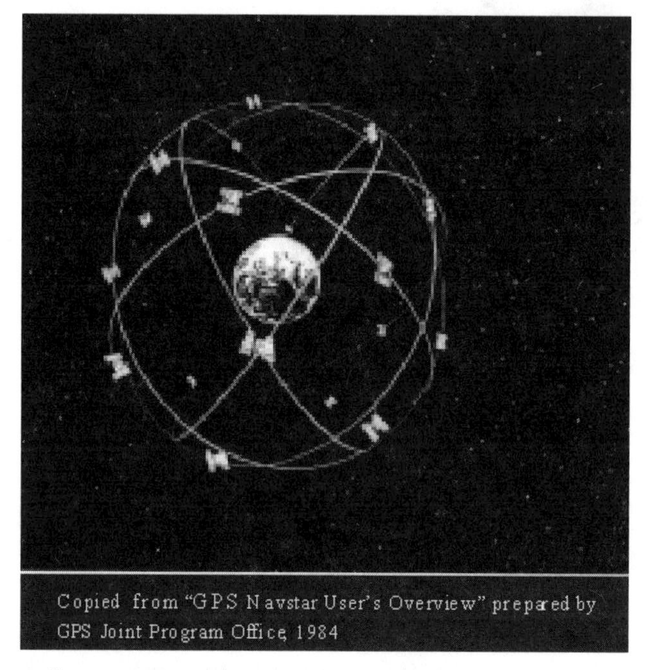

GPS satellites.

Copied from "GPS Navstar User's Overview" prepared by GPS Joint Program Office, 1984

To perform GPS trilateration, you must determine two pieces of information about each satellite: orbital position, and distance from a position on the ground. Orbital position is contained in an "almanac" transmitted to a special GPS receiver on the ground during regular operations. Distance is calculated by multiplying the speed of light (186,000 mi per second) by the time it takes for a GPS signal to reach a ground receiver. Satellite and receiver clocks provide extremely accurate timing. Once this process is completed for four satellites, you have calculated a trilaterated position on the Earth's surface. Locating land positions requires four satellites, but three satellites are acceptable for ocean and sea solutions, where the altitude is always sea level.

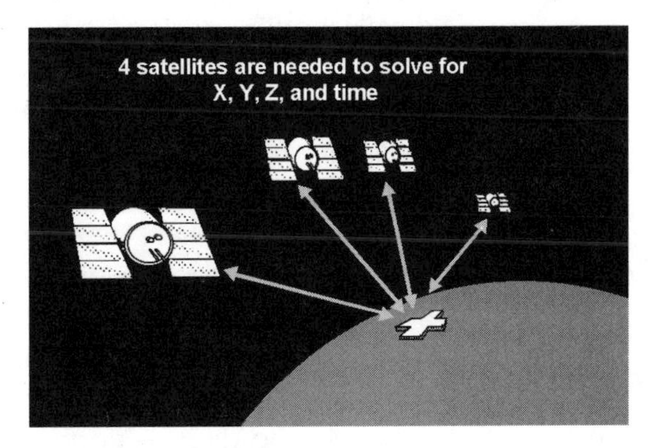

Trilateration using satellite vehicle lines to the ground.

GPS accuracy varies with receiver quality and the correction method you use (described later in this chapter). Survey grade receivers can provide 5mm accuracy, but the normal range is 5mm to 2cm. Resource grade receivers (used in GIS applications) provide sub-meter accuracy with a range of 10cm to 5m. Satellite cost and receiver accuracy are directly proportional; as receiver accuracy increases, so does the cost of using it.

Comparing GPS Data Acquisition to Other Methods

GIS data acquisition and conversion are costly, and data transfer can consume 50 to 70 percent of the total system cost. Even after the initial conversion is complete, maintaining data accuracy is an ongoing task. GPS is a very effective tool for controlling and limiting the costs of acquiring and converting data.

The speed and automated nature of GPS data acquisition can be up to 50 percent less costly than other methods. For example, the City of Ontario, California, compared data collection for a 942-unit fire hydrant inventory using GPS and conventional methods. The GPS method cost $515 and took 41 worker-hours, while the conventional method would have cost $4,575 and taken two to four months. In addition, GPS data offer both spatial and tabular information.

In comparison, digitizing can be labor intensive and result in positional errors, while scanning—although quick—can miss details and require extensive editing. In addition, remote sensing and photogrammetry may not meet the most rigid spatial requirements of GIS, unless high resolution imagery is available. Using GPS, however, spatial and tabular data are collected simultaneously in digital form in the field.

GPS Method of GIS Data Collection and Conversion

GPS data collection begins with the creation of a data dictionary. (See Chapter 15, "Spatial Data Transfer Standards," for more information on data dictionaries.) Using GPS software, links between tabular information and spatial data are defined. The data dictionary defines the structure of point, line, and area features, and attributes and values associated with each feature are also outlined. For example, a well point feature might be described using depth and type attributes, while a road line feature might be described by its name. Attribute values can be input into the GPS data logger unit using menus or text entries. A menu selection for well type might contain the variables domestic, industrial, and agricultural. The road's name would be a text entry. When it is complete, the data dictionary is transferred to a GPS data logger where it will be referenced during the field work.

GPS data collection can be performed 24 hours a day. The next figure illustrates the process for collection, which begins with occupying the defined point, line, or area. "Occupying" in this context means physically visiting the feature in the field and occupying that position while the GPS determines its location. Point features—spatially distributed entities, activities, or events with a single geographic coordinate (e.g., a tree, well, lamppost, manhole cover, blast hole, or fire hydrant)—are stationary GPS activities (see the second figure below). Locating line and area features requires moving along the features in order to collect locational information. As a result, point features have better spatial accuracy than line or area features. As data are gathered, their attributes and values are entered into the data dictionary and linked with spatial location data in the GPS data logger unit.

Using GPS equipment for field data collection.

→ **NOTE:** *In a unique instance of line feature collection, GPS receivers have even been placed on dogs trained in search and rescue efforts. A dog's path (the line feature) is tracked to provide precise directions for rescuers.*

What Can Go Wrong?

While GPS data collection has improved in ease and speed, some obstacles remain. Solid or dense objects can block GPS signals, while wet trees with heavy branches and leaves can mask or attenuate GPS signals as well. Mountains and buildings can block satellite transmission, and multipath signals (or those reflected by nearby land features such as fences) can corrupt GPS data. The resulting delay, as the signals bounce, can spoil measurement accuracy. Advances in GPS electronics have reduced the potential for multipath signals, but GPS field operators are still better off avoiding obvious multipath environments.

Careful planning using GPS mission planning software can reduce the effects of signal blocking by modeling the terrain to display satellite availability. As a result, data collection can be completed at optimal times.

Differential Correction

As mentioned previously, selective availability degrades collected GPS data to 100m, but this deliberate error can be removed using differential correction. Two methods of differential correction are available: real-time and post-processing.

For each method, the process is the same. A base station at a precise known reference position gathers satellite positions simultaneously with a GPS rover (refer to the next illustration). The base station computes an error correction factor by comparing its location with the GPS signal. This factor is then either transmitted in real time to a rover or logged to a file for post-processing. Post-processing differential correction is more accurate and is preferred by GIS users, because it allows more time to perform accurate calculations and can take advantage of better, faster computers.

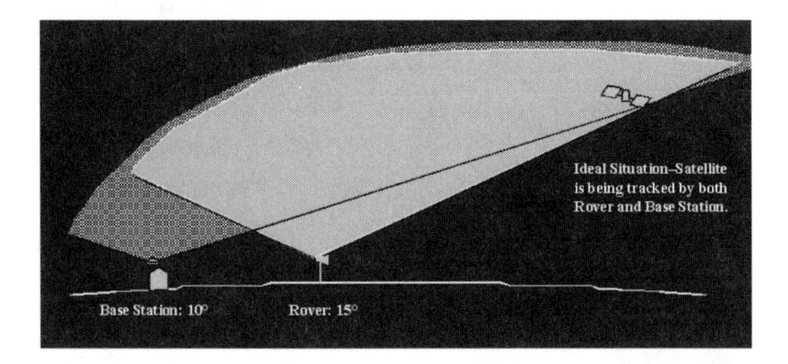
Ideal Situation–Satellite is being tracked by both Rover and Base Station.
Base Station: 10°
Rover: 15°

Base station and rover.

Post-processing differential correction is performed once the data are transferred from the data logger to processing software. The base station correction file is downloaded, and rover and base station data are processed together to complete differential corrections for positions gathered in the field. Positions and feature attributes of the corrected GPS data are cleaned, and then the data are ready to be converted into the proper format for the GIS database. This formatting, usually in the form of a data table, varies slightly according to the GIS software you use.

Data Conversion to GIS

A few years ago, converting data from a GPS to a GIS could be difficult and time-consuming. Today, however, the process has become easy and straightforward. GPS manufacturers provide utilities to automatically convert data and differential corrections into formats suitable for popular GIS software. Data format customization is also possible. For example, many conversions neglect to transfer height because most GIS software is limited to traditional XY coordinates. Customization enables you to transfer GPS height values for contouring or surface treatments.

Other Applications

Most GPS data collection is ground based, but fire management and wildlife inventory, which employ aerial collection, are two notable exceptions. The Bureau of Land Management and the U.S. Forest Service use GPS in helicopters to locate fire perimeters and calculate acreage. In several major fires, aerial and ground GPS data collection has proven invaluable in order to rapidly determine structural damage, environmental damage, and fire extent. GIS is used to analyze and map the fire area. A GPS-equipped airplane or helicopter can quickly locate and inventory wildlife in remote locations or across large areas.

GPS Advantages and Disadvantages

Experienced GPS users learn to optimize data gathering by understanding the system's strengths and weaknesses. Over the past few years, GPS has matured and overcome many shortcomings. For example, before 1993 the full 24-satellite constellation was not in place. As a result, periods existed where four satellites were not available for accurate positioning. This rarely occurs at present. Technology has reduced the potential for multipath errors and GPS data gathering capabilities are growing stronger. However, some problems still remain. The following sections list advantages and disadvantages in using GPS for GIS data collection.

Advantages

- Spatial and tabular data are collected simultaneously.
- Positional accuracy is superior to conventional methods.
- Coordinate systems and reference datums can be easily changed in the field and processing software.
- Conversion into GIS software is relatively straightforward.
- Data collection costs are lower than conventional methods in many applications.
- Visual inspection of features is possible while gathering data.

- Time periods when data can be gathered are virtually unlimited.
- GPS is unaffected by weather.

Disadvantages

- GPS requires training and retraining, as the technology changes.
- "Urban canyon" buildings can block satellite signals.
- Heavy foliage and thick branches can attenuate and/or block satellite signals.
- Multipath reflective signals can make data inaccurate.
- GPS requires careful attention to system configuration and data collection standards and procedures.

The Future

In the future, laser guns attached to GPS receivers will be capable of performing "offsets," enabling operators to reach and calculate the positions of difficult features. They will likely reduce the cost of collecting data and speed data gathering in many cases. For example, a laser "shot" can gather data regarding a manhole cover at a busy traffic intersection without standing directly on (or occupying) it. From one convenient location, nearby utility poles can also be captured with multiple shots. Finally, features on private property can be shot without violating property boundaries.

External sensing devices will also augment GPS data collection; digital cameras comprise the next external sensing wave. Digital images are valuable feature attributes for GIS display and query operations. GPS positions can be joined to scientific data collected by temperature probes and magnetometers. Additional external sensing devices are on the way. GPS equipment will also continue to drop in price, and receivers will become smaller. These trends will increase GPS use for GIS applications. As a result, GPS will likely become a popular and essential tool for GIS data collection and conversion.

Scanning

Adrien Litton

Scanning is perhaps the least understood method of data capture in the field of GIS. Many people recognize the potential benefits of scanning, but few understand the precise nature of those benefits or how to realize them. While scanning is not especially complex, such misconception surrounds scanning that the technology seems shrouded in mystery.

Scanning has existed for some time in other forms largely unrelated to GIS data capture. The mystery arises, in part, from the fact that while it is obvious that GIS scanning bears little resemblance to retail bar code scanning, how GIS scanning differs from engineering scanning is less obvious. Both applications involve large documents that must be displayed on a computer monitor in a digital form. Engineering drawings may even resemble maps. Why aren't maps and engineering drawings treated in the same manner?

Why GIS Data Are Different

The difference lies in the data. Maps and blueprints represent entirely different types of features with very different purposes. The problems you will encounter if you scan maps like blueprints are similar to those you would find trying to make a GIS database out of architectural designs for a house. It can be done, but it probably makes more sense to use a GIS approach for maps, an architectural approach for home design, and an engineering approach to design a flywheel.

The key to successful scanning is to understand the data and how they will be used. Specifically, the key difference between maps and engineering drawings is that maps are spatial and drawings are not. Lines within a drawing may appear to have spatial relationships with one another, but in reality they can be sacrificed for aesthetics or convenience. The "real" information, written next to the lines, must be accurate and precise. Engineers understand that drafted lines are neither accurate nor precise, so they do not rely on them. Instead, engineers rely on numbers appearing next to the lines for the correct dimensions of a part. The lines exist for reference purposes.

GIS observes the reverse philosophy. In GIS, spatial relationships are everything. The fundamental purpose of creating a GIS database is to digitally reproduce spatial relationships. Because the goal of GIS data capture is entirely different from that of engineering, you must take a very different approach when creating GIS features. Text is important, but not nearly as important as lines. Cartographers take great care to ensure that the lines they place on maps are accurate representations of ground features. The discipline of map making has evolved over thousands of years with a very proud tradition of spatial accuracy. It is therefore appropriate that GIS applications maintain spatial accuracy.

As a result, you cannot merely throw a map into a scanner and expect cartographic quality results. GIS scanning is far more complicated. Because industry professionals are interested in the linework, the information must be scanned at higher resolutions. Because the industry aims to maintain spatial relationships, the scanners must be more accurate. Because reproducing cartographic quality on a computer is a goal, the maps must be viewed as they are scanned. Professionals must evaluate each document and make intelligent decisions about the best methods of capturing the data, and they must always undertake scanning with one final objective in mind: to produce a spatially correct GIS database.

Raster and Vector Data

When you consider scanning as a data capture method, you must know what your desired results will be. You must know how you plan to use the data and make decisions regarding the type of output you plan to produce. Raw data from scanners come in a raster format. In a raster image of a stream layer, each pixel would have a value of either 0 or 1. If the value of the pixel is 1, the pixel represents the stream. If the value is 0, the pixel represents something other than the stream. This is known as a bi-level image because it has two values. If you scanned a black and white photograph instead, the pixels would range in value from zero to 255. This model, known as grayscale, assigns a numeric value between zero and 255 to distinct shades of gray; zero represents black, 255 represents white, and all other values are linear shades in between. If you scan a color photo, the scanner assigns pixel values according to the three primary color bands, red, green, and blue.

When a map is scanned, it can be displayed on a computer screen for visual reference. In most applications, being able to view a simple image is sufficient for GIS analysis. The pixels in the image are not spatially related to one another. If you look at the image on the screen you may be able to determine that a given feature is a stream, but the stream pixel itself does not "know" that it is a stream. Nor does it "know" that the pixel to its left is part of the same stream. Consequently, raster data are not referred to as "intelligent" data. You can access images and assign attributes to each pixel, but in simple mapping this is too time-consuming and produces inordinately large data files. In order to make the data "intelligent," you must create vector data from it.

Just as an image represents the original map, vectors represent the raster image. One line in a raster image is made up of a group of unrelated pixels, but a vector is actually the analog definition of that line. Vectors are stored in the form of X and Y point coordinates. They are generally defined as the end of a line (node) or a point where the line changes direction (vertex). A vector is actually just a list of all nodes and vertices that define a raster line. In the CAD world, these are referred to as polylines. Lines with specific, defined shapes (e.g., rectangles or circles) are known as primitives. Because GIS deals with natural and cultural features on the ground, it is primarily concerned with polylines. As such, all raster-to-vector conversions in GIS must create polylines from raster lines.

Advantages of Vector Data

There are several advantages to maintaining vector format data in your GIS database. The main advantage is that vectors are easy to attribute. If your database defines a specific line, it is very easy to add one or more database attributes to the definition. You can define the line as a stream, assign the type of stream, and even name it. When you use your GIS to query on a stream, you can extract all that information from the database. This is generally far more effective than assigning the same attributes to each pixel comprising the stream.

Another advantage of vectors is storage. If you add database attributes to each raster pixel you would eliminate the benefits of raster compression and more than double the size of your image. On the other hand, the size of a vector data file does not vary greatly if attributes are added. In other words, a raster database must have a record for every pixel in the image, while a vector database only requires a record for each line.

Vectors are also easier to update and manipulate because when you select and move a vector, the entire line moves. Raster data are again more problematic because there is no connection among the pixels comprising a feature.

How Scanning Works

Having a basic understanding of scanning technology will help you maximize your scanner's usefulness. There are several different types of scanners, but they all work on the same basic principles. All scanners have a light source, a background, and a lens; they work by assessing the absence or presence of light; and they contain a mechanism for moving the lens, light source, or background past the other two components. The light source can be either fluorescent or incandescent bulbs, the lens can be a regular camera lens or a photosensitive strip, and the background can be the source document or a simple white base.

The simplest scanner to describe and understand is a feed type scanner, and it is also the most efficient scanner for GIS production. A feed type scanner, or direct feed or roller feed scanner, works by moving the document over the camera lenses and the light source, which are fixed. As the document passes through the rollers, a fluorescent light is directed at it through a glass window. The light source is angled in such a way that the light will reflect off the document directly into the camera lens. Beneath each camera lens is a photosensitive device that collects and assesses the intensity of reflected light and sends the information to the computer. This device, called a charged coupled device (CCD), is nothing more than a strip of ceramic with an embedded wire array (or grid) coated with photosensitive paint. When light hits the array, the photosensitive paint establishes an electrical charge, the intensity of which is recorded as a pixel value. The theory behind the CCD is that light reflected off of dark material will have a lower charge than light reflected off lighter material. In a grayscale scan, this value is recorded directly, although it may be adjusted for contrast and brightness. In a bi-level scan, grayscale values are translated to 0s or 1s.

Color scanners work somewhat differently, but the principle is the same. Light intensities are collected by the CCD for each color by filtering the lens and performing multiple passes or using a special CCD divided into three portions for sensing red, green, and blue. The latter method is more efficient and allows color scanning to work on a roller feed scanner. Color scanners are addressed later in this chapter.

About Scanners

Several different types of scanners exist for various purposes. Scanning has been used in medical applications, inventory control, data archiving and retrieval systems, document management, and engineering since the 1970s. While scanning was introduced to the GIS industry in the mid-1980s, it only became an acceptable means of data capture several years later.

When scanning was first used in GIS, professionals incorrectly assumed that scanning worked in the same manner in all industries. It took a great deal of effort and failure before the industry learned how different GIS really is. After experimenting with existing applications, the GIS industry eventually learned that it had to develop a new scanning model specific to GIS. To do so, it had to learn far more about scanners themselves.

- **Desktop scanner.** Often referred to as flatbed scanners, desktop scanners are the most common type of scanner today. They can accommodate standard letter size paper and slightly larger documents. The document is placed face down on a plate of glass, and the light and camera (a single unit) move on a track beneath the glass. A lid covers the document to keep out excess light. Desktop scanners are the primary method of data entry for document management. They are also used in desktop publishing systems, reprographics, graphic arts, and computer animation applications.
- **Drum Scanner.** The first large format scanners developed were drum scanners. Reminiscent of old mimeograph machines, drum scanners operate by affixing a document to a large cylinder (or drum) that spins around while the scan head (light and camera assembly) moves from side to side capturing data. Drum scanners remain widely used in photogrammetry, medical applications, or other types of data capture requiring extreme precision.

- **Large format feed scanner**. As mentioned previously, the most common scanner used in GIS production is the large format feed scanner. It works by moving the document through a roller assembly over the cameras. This is the only scanner that moves the document rather than the scan head. The cameras are fixed in place, and the scanner's accuracy depends upon consistent camera mounting and the rollers' ability to move the document evenly. Feed scanners are primarily used in engineering applications, GIS, and applications involving large documents, but not requiring extreme precision.

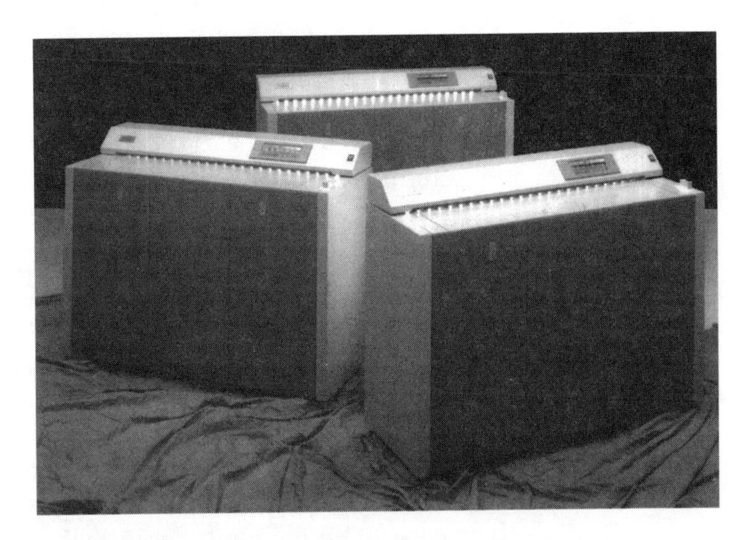

Large format feed scanners.

Which Scanner Is Right

You must understand the strengths and weaknesses of each type of scanner in order to determine the correct one for your applications. Desktop scanners are not well suited for documents of varying quality or different types of data. Moreove, they are impractical for large map sheets.

Drum scanners are extremely accurate and can accommodate large documents, but they are extremely expensive and relatively slow; scans can take 15 to 20 minutes. However, they are ideal for creating digital orthophotography.

Consequently, feed scanners are the primary tools for capturing data on large documents in digital form. Feed scanners can usually produce high quality images in less than five minutes and their cost is generally within the budget of GIS data conversion firms with many maps to scan.

Feed scanner drawbacks include a lower accuracy than desktop or drum scanners. You must also take care when scanning old, fragile documents because the rollers can damage the map. Therefore, while GIS scanning probably entails choosing a feed scanner, be selective. Take the quality of your documents into account, and understand precisely the accuracy requirements of your GIS database.

Image Quality

The usefulness of scanning to produce high quality raster images is apparent to most GIS users, but have you really determined what you want out of your images? Obviously, the scanned image should be a reasonably accurate representation of the original, but there are other issues to consider. You must know how the images will be used before you can determine how they should be scanned, in order to achieve the best possible output for your applications. Are you archiving old maps so that they can be reprinted later? Are you creating an accurate GIS landbase from drafted layers? Are you fashioning a high quality visual backdrop for GIS data display? Understanding your application requirements will help you determine the best way to prepare and scan your map, and save you headaches during production.

Digital versus Visual Output

There are two primary outcomes of scanning in GIS: visual and digital output. If your goal is visual output, you want to create an aesthetically pleasing raster image. This includes being able to distinguish various types of linework, read the text, and identify map symbology. Ground features should be visible on a scanned photograph. Because the scanned image is the final product, it should be as visually perfect as possible.

On the other hand, when you scan for digital output, the raster from the scan will likely be converted into vector form. If you are investing time and resources to scan for vector output, it may be wise to automate the process as much as possible. This means creating an image that a computer will be able to interpret (and sacrifice some of the aesthetic attributes of the scan). The key to creating the best scan for vectorization is realizing that a computer interprets a raster image differently than the human eye. When scanning for vectors, you are really scanning for the computer. Linework must be sharp and clear. Text and other features not being vectorized must either be ignored or their impact reduced so that they will not interfere with the vectorization process. If done correctly, the final image you produce may not look pretty, but your vectorization software will love it.

Thresholding

How do scanners control output in order to create an aesthetic image or an image appropriate for vectorization? In bi-level scanning of line art, the key is thresholding. Remember that bi-level scanners are really just grayscale scanners with added options. The scanner sees data as values ranging from zero to 255. Thresholding is the process of the scanner picking a point in the grayscale above which all values are translated as pure white (255) and below which all values are translated as pure black (zero). This threshold is where the majority of the data lie. The result is a bi-level raster image. Setting the correct threshold value allows you to capture the best possible cross section of valid data.

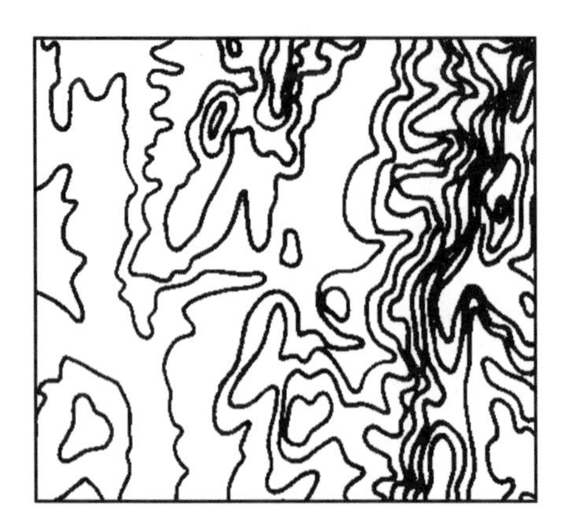

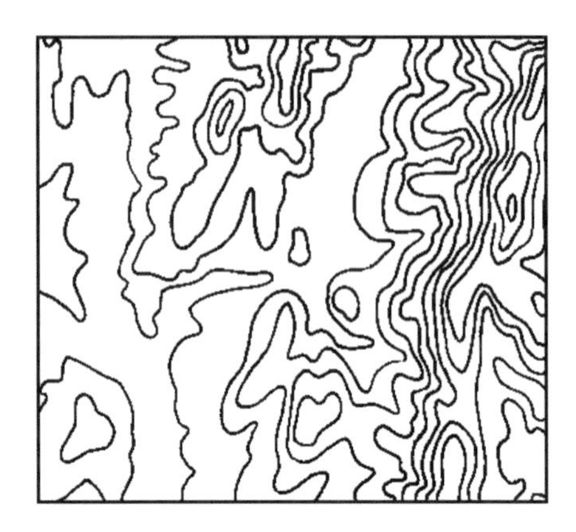

A bi-level scanned image with a low threshold setting (top left). A bi-level scanned image with a medium threshold setting (top right). A bi-level scanned image with a high threshold setting (right).

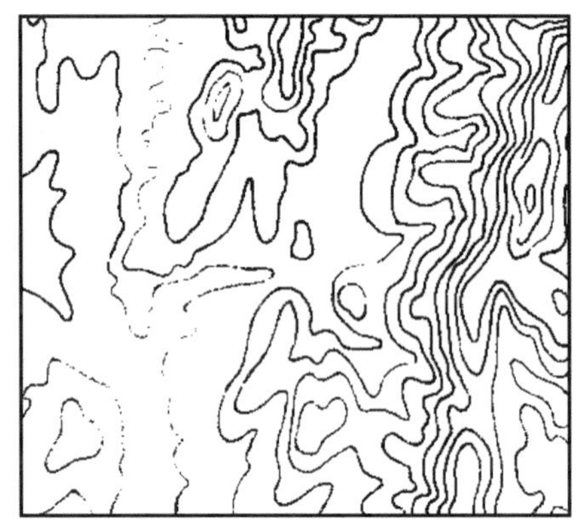

Brightness and Contrast

Threshold values are inappropriate for grayscale images because the best output will retain 256 values. Scanners tend to see things differently than the human eye, but computer display screens are designed to display data so that you can readily interpret them. A scanner sees absolute values, but the monitor displays them as averages. This discrepancy creates images that are often too dark to interpret on a standard monitor. Some image viewing programs correct this phenomenon by adjusting the picture you see, but you may be better off adjusting the image itself. You can adjust the image the same way you might on a black and white television screen, using brightness and contrast.

Depending on the image, all 256 shades of grayscale will probably not be used. In addition, other values can be grouped into a single value because you would never be able to distinguish them anyway. Most monitors group absolute grayscale values toward the bottom of the band, because scanners see darker values than the human eye, and true white does not exist for scanners. The easiest way to brighten an image is to move all values slightly up the grayscale. Because an average scan of an aerial photograph only contains approximately 75 values, there is ample room in the grayscale band to brighten the image. Many grayscale scanner interfaces have a brightness adjustment that instructs the scanner to convert the lowest value to pure black. This helps produce adequate image brightness.

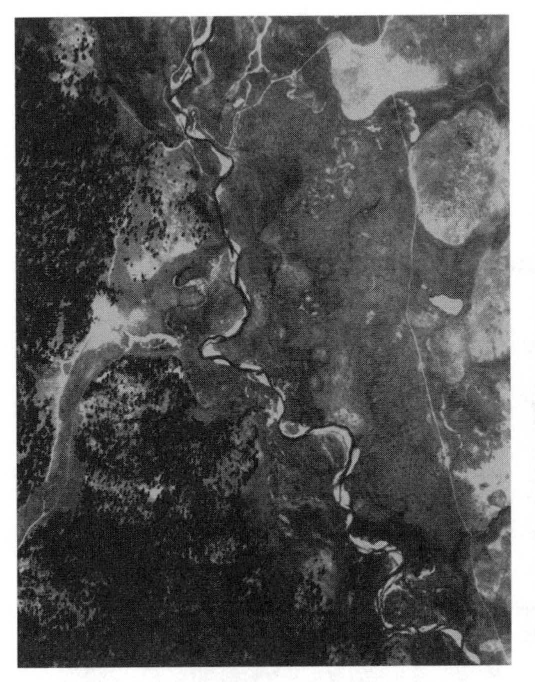

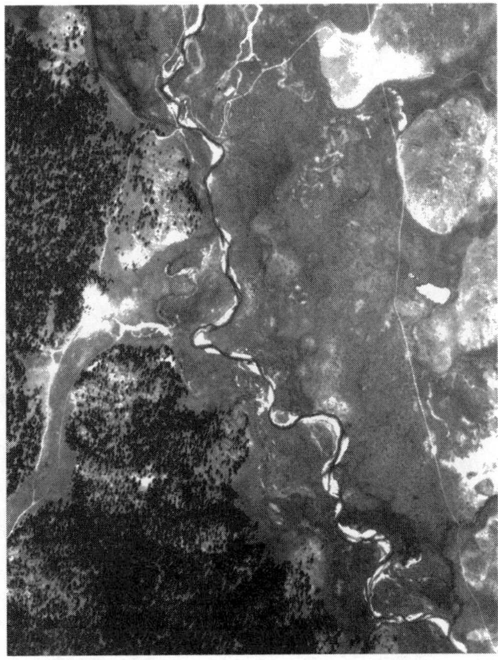

A scanned photograph with no brightness adjustment (left), and a scanned photograph with adequate brightness adjustment (right).

Grayscale images may also contain values too close for the human eye to distinguish. This is especially evident in aerial photographs where water and densely forested areas meet: it is difficult to tell where the water ends and the trees begin. Because the purpose of a scanned aerial photo is to distinguish between these types of features, such a scan is virtually useless. The solution to this problem is adjusting contrast. Contrast is merely stretching the 75 values contained in the image across the grayscale band. If all dark areas are separated by only one or two values, they become concentrated and blend together. However, if you separate them by five or six values you can see different tones. Adjusting the contrast allows you to determine the correct number of values to place between each pixel value, so that the image is easier to interpret.

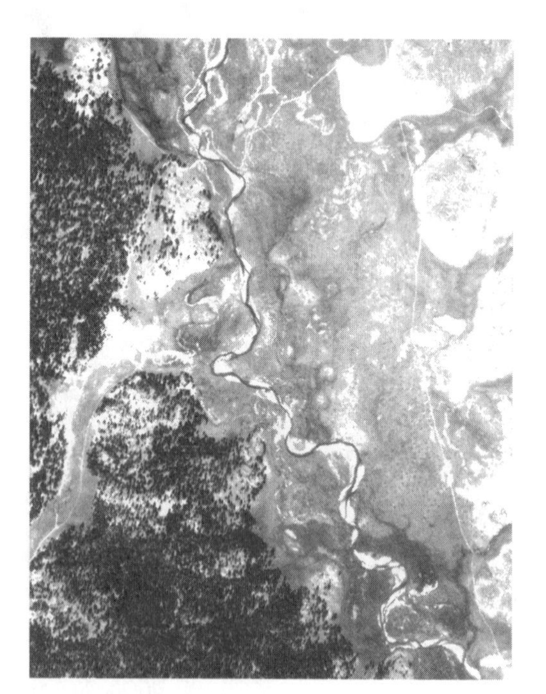

A scanned photograph with low contrast setting (left). A scanned photograph with high contrast setting (right).

Gamma Correction and Thresholding

Brightness and contrast corrections will enhance most photographs enough to yield aesthetic images that can be easily interpreted. However, they are limited to situations where differences are consistent throughout the document. But what if differences are inconsistent? For example, it is possible for a single image to have extremely dark areas falling at the bottom of the grayscale band and light areas falling near the top, such as a forest bordering a beach. While the forest and ocean will be dark, the beach will be extremely light, and adjusting brightness will only make the bright areas more difficult to interpret—and not very aesthetic, either. Using contrast adjustment will eliminate values at the extremes of the band. Such situations require different corrections for the lighter and darker areas. Otherwise you will sacrifice one or more feature types.

The method used to fix this problem is known as gamma correction. In this process, the scanner analyzes a histogram of the image (simply a count of all cells in the image along the grayscale) and places points strategically along the histogram to isolate data types. This allows the image processor in the scanner or the scanner interface software to automatically select the best areas to brighten, darken, or stretch the contrast. Gamma values are usually selected along a logarithmic scale between 1 and 2. A value of 1.2 is twice the correction as a value of 1.1. Gamma scales vary among scanners and may have different terms, but the concepts are similar and the methods for image enhancement are usually well documented in the scanner's user guide.

Dynamic Thresholding

Another type of image enhancement, dynamic thresholding, is used to clean "noisy" or speckled documents, such as old blueprints. A blueprint is paper on which a white emulsion is placed on top of a blue emulsion. Where lines are drawn, the blue emulsion shows through. This is similar to self-carboning paper. However, as the document ages, the white emulsion begins to wear off and more blue emulsion shows through. This creates speckles, lines, and other marks that the scanner might mistake for data. Because the noise did not exist when the blueprint was created, it tends to be lighter than the original data, but it can become so problematic that it eventually obscures valid data. Modern scanners can enhance the lines and suppress the noise so that they produce an image even better than the original map.

In dynamic thresholding, a filter is passed over the image as it is scanned and all pixel values are evaluated. Pixels darker than their neighbors are translated into data, and lighter pixels are suppressed. Dynamic thresholding usually involves maximum and minimum values for valid data, so that erroneous data are filtered out. The process in general works quite well, but take care to ensure that valid data lines are eliminated as well.

Throughput

Scanned image files tend to be very large. Compared with vector data, raster files can be enormous, and many users are unprepared for the immense volume of data created in GIS data conversion. This can lead to serious storage and data transfer problems. At present, nearly all scanners are connected to SCSI ports with a special board installed in the computer to receive and process data as quickly as possible but problems may still exist. Hard disks cannot always keep pace with the data as they are transferred through the SCSI port, and SCSI technology tells the SCSI input device to wait while the output device catches up. This means that the scanner temporarily stops sending data, and you will receive data buffer overflow errors. Other scanners will simply halt the scan until the computer is ready to receive data again, and then resume scanning. While this may seem like a helpful feature, when the feed mechanism must stop and restart scanning a map, the document may slip and introduce spatial inaccuracies into the data.

Optimal Environment

The best way to avoid such problems is to ensure that you have an optimal environment for scanning. Your hardware platform should reflect the state of the art technology at the time the *scanner* was new. As scanners evolve, they take advantage of changes in computing. Processors become faster and scanners become smarter. You may run a 1990 scanner on an old 386 PC quite well, but when you upgrade the scanner, you are certain to run into problems if you do not upgrade your PC as well. Keep in mind that scanning is machine intensive, and GIS data capture calls for as much processing power as possible. Consequently, do not settle for the minimum configuration.

Memory Guidelines

You can also prevent problems with large scans by adding memory. Virtually every hardware device will specify minimum memory requirements. When dealing with large images, you cannot settle for minimums. It is much faster for an input device to write to memory than to a disk because there is no physical transaction involved. The scanner's installation guide may specify 16 Mb of RAM, but that may be misleading. Instead, think about the size of the images you will create. If you are scanning large photographs or specifying high resolutions, your images will likely exceed 16 Mb—they may exceed 80 Mb. The most efficient way to scan is to bring the entire image into memory, and then move it to a disk. This requires having enough RAM to contain the entire image. While such a setup may not always be practical, aim for as much memory as you can. For normal scanning you should have a minimum of 16 Mb, but 32 Mb is better. If you are scanning large images, you should have 64 Mb or more. If your scanner host computer is going to be performing other tasks in tandem with scanning, double your numbers.

Another issue that affects throughput is your ability to quickly and accurately review image data. This requires a high resolution monitor that can accommodate your image. Scanner software documentation will usually specify recommended monitors and video boards. Read this information carefully. If you cannot determine a scan's quality by merely displaying it on the screen, you may spend countless hours rescanning maps.

Resolution

Resolution is defined as the density of a raster image. An image is made up of dots, or pixels. Resolution is the number of pixels in a given unit of measure. The most common unit of measure in scanning is inches. Therefore, the resolution of an image is usually expressed in dots per inch (dpi). Keep in mind that this is a linear inch, however, not a square inch, so scan resolution is a measure of linear density, not area.

Scanner documentation often uses the term "scan line." A scan line represents a single pixel row of an image. It is one pixel in length and follows the entire document width as it is read into the scanner. Scan lines are built by the scan head based on the width of the scanner array. If the scanner has a moving head, the array reads a portion of the scan line and then moves into position to read the next portion of the scan line, and so on until the entire scan line has been read. Once the first set of scan lines is complete, the scanner head resets itself to read the next set. On drum scanners and large flat bed scanners, which have moving heads, the process of scanning and building scan lines is tedious and time-consuming. This is the primary reason that feed scanners tend to be so much more efficient than drum scanners. In feed scanners, the scan head is fixed in position and the document moves across the scan bed, which enables the scanner to read as many as five scan lines simultaneously.

Scan Heads

Scanner resolution, of course, is determined completely by the scan head, and the type of scan head determines how resolution is computed. The two most common types of scan heads are the moving scan head, typically used in drum scanners, and the fixed CCD array, used in most feed scanners.

NOTE: *The single strip array is a new type of scan head gaining popularity. It relies strictly on sensitivity to continuous direct light. Because a single array strip reads continuous light, there is no limit to the amount of resolution obtained. Still in its infancy, this technology requires intensive user testing to prove its viability.*

Optical versus Interpolated Resolution

Optical resolution is what the scanner is physically capable of seeing. Because most feed scanners use 5,000 element CCDs, you can use a simple formula as a rule of thumb. Take the number of cameras in the scanner and add one. Then multiply by 100. This will give you the optical resolution of the scanner. For example, if the scanner has three cameras, dpi is 400. This formula works in most cases. Beware of manufacturers' claims that boast 1,000 dpi if the scanner has only eight cameras. Look closely at the scanner specifications. It will probably state that it has a 5,280 element CCD, in which case it is capable of slightly higher resolutions with fewer cameras. Manufacturers use interpolation to increase resolution, but this may actually backfire and lower image quality.

Interpolation

There is a common misconception that higher resolutions produce better output. Scanners sound more impressive to consumers if they seem capable of scanning at higher resolutions. However, interpolated resolutions that double scanning dpi require a fairly sophisticated algorithm capable of approximating higher resolutions by adding pixels to the output. The complexity of this algorithm lies in the fact that you cannot simply append pixels with arbitrary values to the output data. You have to append pixels with the correct values to the output data.

Essentially, if neighboring pixels produce values that increase or decrease in a linear fashion, it is easy to determine the correct pixel value for the appended pixel, but if the values are erratic, the scanner will have no reliable method for knowing which value to place in the appended pixel. Interpolated resolution works very well with grayscale images because pixel values almost always change in primarily a linear fashion. The result of interpolated grayscale is an image that degrades less when zoomed. However, interpolated resolution is not well suited to determining edges. This means that the analysis of neighboring pixels can often become erratic on linear data.

Interpolation does not work well with line art or bi-level scans, either. At best, it will produce "ghosting" of the line, or a line that is fatter on one edge than on the other. This will consistently offset the line center in vectorization by half a pixel if the line is straight and all lines are far apart. However, as the direction of the line changes, so does the ghosting effect. In simple cases, inaccuracies of half a pixel are not substantial problems and can usually be absorbed into the overall accuracy of the database. However, in simple cases, you rarely require more than 400 dpi anyway. Where you might want 800 dpi resolution—with complex contours, for example—you would require high resolution in order to separate lines as clearly as possible and obtain the most pixels in each line for vectorization. Your goal in such cases is to scan at 800 or 1,000 dpi.

Remember that the scanner only sees 400 dpi. If the space between lines is less than 0.0025" wide, the scanner cannot see the space. Also remember that the scanner is reading and interpreting data along a horizontal scan line. In order to interpolate a value of data or no data, the scanner must look at adjacent pixels at 400 dpi. In complex contours, it is very likely for an "off" pixel (blank space) to be surrounded by "on" pixels (data lines) on either side. The interpolation algorithm will calculate the interpolated pixel value incorrectly as "on" instead of "off." This means that a space between contour lines could be written as a false contour line.

The same problem can be applied to an "on" pixel value between two "off" pixel values. The scanner would interpolate a valid contour line as blank space, resulting in missing data. Consider how irregular contour lines seem when viewed five scan lines at a time. In tight places, no interpolation algorithm can accurately write out contour lines. The result is that some contour lines will be captured accurately and others will be offset by half a contour value. This means that valid lines are omitted, while erroneous lines are placed adjacent to the locations where the valid lines were supposed to be.

To make matters worse, there is no way to tell where in the document these errors will occur and no way to repair them. Creating a digital terrain model from an interpolated scan would produce dangerously inaccurate results. As such, you must be very cognizant of interpolated resolution. A resolution of 400 dpi is adequate for most scans, but with most scanners, there is virtually no interpolation at 400 dpi. Where higher resolutions are indicated, it is extremely important that resolution be achieved optically, not through interpolation.

Choosing the Right Resolution

Now that you are familiar with resolution and how to determine the resolution capabilities of the scanner, how do you determine your own resolution requirements? This process is not very complex. If you are scanning simply to view the data on screen, you probably will not require very high resolution. However, zooming in will require higher resolution. For image backdrops, you can determine this easily through experimentation.

NOTE: *Scanning at twice the resolution will result in a file four times as large.*

Scan a document at various resolutions and see what works best for your application, but remember that higher resolutions produce larger file sizes. This will increase storage requirements and drawing times. A high resolution image may produce a better aesthetic result, but image quality may not offset overhead costs. If you must zoom in very closely to some areas, it may be appropriate to scan the entire document at a lower resolution for full view display, and then rescan it at a higher resolution, cutting out the pieces you will view in detail.

✓ **TIP:** *A good rule of thumb is to estimate that you will rarely require more than 200 dpi to view text and line art, or color lithographs; 300 dpi to view most photographs; and 400 dpi to view high quality orthophotos.*

When scanning for aesthetics, the best way to determine resolution is trial and error. When scanning for vectors, determining appropriate resolution can be more scientific, but is still relatively straightforward. Most vectorization programs require that lines be at least three pixels wide in order to be reproduced as a vector. A typical hand drafted line in pencil is approximately 0.018". In order to vectorize a line of this width, you might be able to use 200 dpi scans, but 300 dpi is more reliable. This line size is common for plats, tax maps, and typical blueprints. However, you may need to capture manuscript lines with more care. A line drafted in ink or fine point pencil will be about 0.012" wide and require approximately 400 dpi scanning to vectorize properly. The U.S. Geological Survey standard for linear features is 0.006" and requires at least 500 dpi. Scribed lines tend to be about 0.003" wide and should be scanned at 1,000 dpi.

You may have noticed the apparent correlation between drafting quality and resolution. The higher the quality of the source, the higher the resolution necessary to accurately capture it. If you are scanning extremely accurate scribed plates, you will want to preserve the quality of the data, so higher resolution is necessary. However, if you are scanning hastily drafted plats or old blue lines, lower resolution is appropriate. In fact, choosing the correct resolution is a useful preliminary step for producing the best possible data.

In today's demanding world, there is a tendency to believe that the highest resolutions will produce the best output. This assumption could not be further from the truth. The resolution you choose must be appropriate to the data you plan to capture. For example, if you are scanning a fairly old hand drafted map, the pencil will be faded and small holes will likely exist in the linework. When the scanner sees the holes, it will try to reproduce them exactly. If you scan such a document at 1,000 dpi, you would get a very accurate representation of each error in the linework, and the errors likely would be reproduced accurately in the vectorization process as well. However, you are scanning to capture linework as intended by the cartographer, not linework errors.

Scanners and vectorization packages, like everything else in the computing world, tend to be very literal. If you feed data with holes into the scanner, you are likely to receive data with holes in your output vectors. The solution to this problem is to reduce the scanner's optical resolution by turning off some elements in the CCD. This will make the pixels larger and redistribute light intensity. This maneuver will "blur" many holes into valid data. Therefore, sometimes lower resolutions produce better final results.

You have probably noticed that in scanning, a little math goes a long way. Strong math and logic skills are a necessity for understanding resolution and how different resolutions affect data. In most cases, 400 dpi is the most appropriate resolution for scanning, and because most scanners have a maximum optical resolution of 400 dpi anyway, this is the resolution you are likely to encounter. If you anticipate having a frequent need to scan at higher resolutions, you should procure a high resolution scanner. You will probably never have a need in GIS applications to scan at higher than 1,000 dpi. Even with a high resolution scanner, it is appropriate to scan most map data at lower resolutions.

While 400 dpi works in most cases, you may want to consider scanning images you vectorize at 500 dpi if your scanner is capable of 1,000 dpi. The reason lies in math. There is virtually no cost difference between scanning at 500 dpi versus 400 dpi. A 400 dpi image is smaller and processes faster, but a 500 dpi image has slightly higher quality. These attributes cancel each other out. However, 500 dpi is a far more convenient resolution because the math is easier. At 500 dpi a pixel is 0.002". At 400 dpi a pixel is 0.0025". This means that calculations will have to be carried out an extra decimal place and may give odd-numbered answers. Therefore, 500 dpi is better for the simple reason that you can do most of the math in your head.

Scale and Accuracy in Relation to Resolution

Map scale and image accuracy often become confused in the scanning process. In photogrammetry and other digital imaging applications, precision and accuracy are directly controlled by image resolution because resolution is typically measured in ground units. This assumes that the image is already in accurate ground coordinates. You must remember that scanned images are created from hardcopy maps and photos and as such must always be processed at a scale of 1:1 within the dimensions of the original document. You can reproduce an orthophoto in digital form, but you cannot create one through scanning alone. Uncorrected photos, once scanned and georeferenced, remain uncorrected photos. A scanned and georeferenced orthophoto will not have the same precision as a digital photo created with a stereoscope.

If accurate ground resolution is a requirement of your GIS, you will need to purchase digital orthophoto data or create them with proper photogrammetric procedures. Consequently, when scanning for GIS data capture, scale and accuracy have no real tie to resolution. High resolution will only control the density of the output image; it will not directly control precision or accuracy. Precision is determined by the scale of the original document, and accuracy is controlled by the stability of the media and the scanner's physical accuracy. The document you choose to scan and the quality of the scanning equipment will equally affect the quality of your output.

Scanning Accuracy

Accuracy is paramount in GIS. Because GIS relies on spatial relationships, you must know precisely where on the ground a given map feature lies. Thus, when capturing GIS data, you must capture all features as accurately as possible. If accuracy is not controlled by resolution, how can you ensure data are being captured accurately?

What Accuracy Is Required?

First, determine the accuracy requirements for your GIS. While you may lean toward having the most accurate data possible, avoid this temptation in the context of scanning. Quantify your accuracy needs correctly before you purchase lavish, expensive equipment or make incorrect accuracy assessments.

Do not exhaust resources trying to attain a higher accuracy than exists in your source material, and avoid adding inaccurate data to an otherwise accurate database by attempting to cut costs on scanner equipment. Remember that the accuracy of your database hinges on the accuracy of your data. If you are scanning maps with no true scale or projection, they are only estimates of ground positions. It is pointless to measure the scanning accuracy for such maps minutely, because accuracy has no relevance. Conversely, if the source map is a highly accurate rendering of ground features drafted to extreme cartographic precision, you should be as precise as possible when measuring the accuracy of the output.

Accuracy and Source Documents

One way to control the accuracy of your scan is to ensure that you have the best source available for scanning. If accuracy is important, attempt to find a stable media to scan. Paper maps are not considered stable. A paper map can stretch or shrink hundredths of an inch during the course of a day. Mylar is a much more stable media, but understanding Mylar weights is also important. Mylar weight is measured in mils (thousandths of an inch), and typically, it comes in three weights. Four mil Mylar, often referred to as vellum, exhibits the same instability as paper and can tear easily in the rollers. If you must choose between high quality paper or vellum, use paper. Eight mil Mylar, the thickest, is the most stable, but it can drag in the rollers of a scanner. Generally, your best option is six mil Mylar. It is far more stable than paper and will not drag in most rollers.

Even if you scan the most accurate source available on optimum media, the document is still subject to environmental conditions. Extreme temperature or humidity changes will change the shape of even the most stable media. As a result, some GIS professionals attempt to precisely control the environment in which scanning occurs. Unfortunately, this may cause more problems than it solves. Imagine a typical plotter room. The temperature is maintained at a constant 72° F with a constant humidity of 60 percent or less. This ensures that plots do not change shape while they are being created. As soon as the plots leave the room, however, they are subject to the same environmental hazards as any other map. Scanners are often confused with plotters, and some people assume that they must be operated under the same conditions. This is a dangerous assumption because the accuracy of the scan depends on having correct environmental conditions.

The rule is to scan your documents under the same environmental conditions as the source documents are normally exposed to. This means that your scanner can and should be operated in the same environment where you store and work with the original documents. Because scanner operation is based on light, and light is nearly always accompanied by heat, your document will be slightly affected each time you scan. This temperature change (of one or two degrees), will usually not affect the output image. If you take a document normally exposed to temperatures ranging from 70° F to 85° F and put it into a controlled environment, you are imposing dramatic environmental change. Basically, your document will be a different shape when it leaves the controlled environment than when it entered. This means that the scanned image no longer reflects the true accuracy of the source document. However, if you scan the document in the environment the map is normally accustomed to, the output image will have the same shape—and therefore accuracy—as the source.

Scanner Accuracy

Controlling the accuracy of the source is only the first part of the process. Your scanner must also be accurate enough to meet the requirements of your GIS. How do you know if it does? Again, you must accurately assess your needs. If your GIS project has no predefined accuracy standard, you must determine a standard for scanning. One way to do this is to measure the width of a typical line on a source document. If a line on the scanned image is smaller throughout the document, then it is generally considered accurate enough to incorporate into your GIS, unless more stringent standards are specified. Most hand drafted lines are 0.020" wide, and most scribed lines are 0.006" wide. To ensure that the width of a vector line falls within a drafted line width on the original document, it must be accurate to 0.018" for most drafting and 0.005" for scribed plates. Blueprints, facility drawings, and maps without scale can have accuracies as little as 0.025".

How can you determine what accuracy level your scanner is capable of? For most types of hardware, specifications such as accuracy are generally listed in the published design specs of the device. Marketing brochures for virtually all large document scanners will provide accuracy information in the form of a percentage. For instance, the document may say that the scanner is accurate to ± 0.04%. This is so common it can be assumed as standard. According to this standard, nearly all document scanners have the same accuracy. However, years of experimentation and evaluation have revealed that this is not the case. Some feed scanners are more accurate than others, yet they continue to publish the same accuracy specifications. To understand why, you must understand what the manufacturers' accuracy specifications really mean.

Manufacturers provide scanner accuracy in terms of a percentage, but you must quantify it as an absolute value. In order to convert a percentage into a number, you must know how the number was derived. What the specification means (but is rarely documented) is that accuracy is determined as a percentage of scale across the entire dimension of the document. In other words, if the source document is 36" long, the resultant image will be within ± 0.04% of 36" when measured at the scale of the source. That is, the output image will have an overall length from 35.9856 to 36.0144". If you carry this logic out, this information will lead you to the conclusion that the absolute accuracy of the scanner is 0.0144" averaged over a 36" document. Based on the manufacturers' specifications, nearly all scanners appear to have an absolute accuracy of 0.0004".

This seems to be more than accurate for the requirements of any GIS. It also makes feed scanners appear to be more accurate than drum scanners boasting an absolute accuracy of 0.0005" (50 microns). But you already know that it is impossible for a roller feed to be more accurate than a fixed drum. The answer to this discrepancy lies in the manufacturers' definition of accuracy. According to scanner manufacturers, an accurate image is one with output dimensions proportional to its source. The output dimensions can be controlled by software. For example, if a scanner produces an image 36.5" instead of 36", the image can be automatically rescaled to 36", and a 1/2" error was redistributed throughout the document. The error was averaged without consideration for the features within the document. The result could be a perfectly proportioned image with gross inaccuracies throughout.

To summarize, manufacturer specifications provide no reliable indication of a scanner's true accuracy. In order to be truly accurate in GIS, you must know a feature's ground position. To scan accurately, you must be able to control where the feature will fall within the document, and adjusting the document's scale will not suffice. You must control the accuracy of the image as it is being created. While minor errors can be averaged out through the automated scaling process, large errors will only be amplified.

If you have concerns regarding your scanner's accuracy, it may be wise to seek advice from a scanning professional.

Map Preparation for Scanning

Once you have calculated the correct resolution for scanning and determined that the scanner will meet your accuracy needs, you must prep the map. Some maps can produce useful output without prepping, but others require detailed preparation. Keep in mind how you intend to use the scan for your GIS.

Eventually, you must put the data in georeferenced coordinates. The easiest way to accomplish this is using control points. On drafted maps, you must place drafted markings known as tic marks with known geographic locations. If you intend to vectorize, you must also prep the map for this specific step. Know the limits of your vectorization software and prep accordingly. Lines are often represented symbolically on maps. Because most vectorization packages cannot handle symbology well, symbolized features must be filled in with ink. You will frequently encounter polygon features that must be outlined, but the map feature may be a vignette and provide no clear edge for vectorization. In such a situation, you must draft in the edge of the polygon and fill in spotty data.

Dashed or double lines may require work as well. It all depends on your data. Because map prepping is unique to each GIS application, it is impossible to provide clear, detailed instructions regarding how and what to prep. You must develop your own requirements. However, mentioning several costly pitfalls may save you from trouble.

- Never scan a map without control points.

Always ensure that you have correct and accurate registration marks on your document. These should be drafted in ink with a 0.003" liquid ink pen. Your tics should be the highest quality data on your map. Do not rely on photo registration marks because they may have no tie to ground positions. Tic marks should always have specific, known geographic locations. Avoid stick-on registration marks because they can move during scanning and therefore be useless.

- Never scan a dirty map.

Maps are often stored in bundles with packaging tape; this is especially true of Mylar separates. Because the weight of the Mylar can cause rolls of separates to unravel, very sticky tape with cloth fibers is frequently used to ship the separates. This tape can leave a gum residue that will melt and smear on the document during scanning, creating erroneous data in your output image. It may also damage the glass on your scanner. Always clean Mylar separates with photo cleaner if they have been exposed to tape before scanning them.

- If you use tape on a map, use only the best tape available.

Tape residue can melt during a scan, and tape itself can move. If you want to affix additional data to your scan, use only the best tape you can find. Never use regular masking tape or packaging tape. Drafting tape (a type of masking tape) is acceptable, because it leaves no residue and does not move easily in the rollers. If you use transparent tape, use only Scotch Brand Invisible Tape.

- Never use petroleum based permanent markers.

When filling in map features, always avoid oil based magic markers or laundry markers, because they can destroy your scanner. Petroleum based inks used in felt tip markers will bond with the scanner glass as it heats. Because glass is porous, as the ink melts it will be absorbed into the glass, leaving a permanent mark that cannot be removed. Your only option is to replace the glass in the scanner to eliminate erroneous data left by the ink residue. Opt for markers with water based inks.

- Never draft with graphite.

Most pencils are made of graphite, and they work by leaving deposits of faceted material on a document. The facets reflect ambient light, allowing you to see the mark. However, when you shine a light directly at a graphite line, the facets may refract the light away from the line, rendering it virtually invisible. Because the scanner depends upon reflected light, it will be unable to see every line, and your results will suffer. If you choose pencils (because they are faster and easier than ink), use wax lead pencils found at drafting supply stores.

- Avoid glossy finish photo reproductions.

When obtaining photographs for scanning, specify matte finish. Glossy prints will refract light and produce a poor scan. Aerial photos and orthophotos are usually produced on matte finish paper, but reproductions are typically printed with a glossy finish, unless you request otherwise.

Vectorization

Once you have decided to translate your scanned data to vector form, how can you convert them? Because a scanner produces raster images, you will need vectorization software. The simplest way to create vectors is to digitize them.

You can do this on screen using your raster image as a backdrop in a CAD package (heads-up digitizing), but this process can be slow and tedious. A more automated approach is usually faster and tends to produce more consistent results. If you let the computer generate vectors it will interpret all lines in the same manner and reduce the inconsistencies you might produce through manual digitizing. There are two basic approaches to automated vectorization: batch and interactive mode.

Batch Mode

Batch vectorization takes the entire raster document and converts all elements to vector format simultaneously. You can stack several batch jobs together and run them sequentially, which allows you to scan more documents while the first set is vectorizing. If you work efficiently—the key to batch mode—you can capture a large amount of data in a very short time with this process. Depending on the size of the raster image and the vector package used, it can take 30 seconds to 15 hours to batch vectorize a map layer. When the vectorization is complete, you must edit the data to remove errors created during the process and then create attributes. Batch vectorization is generally more appropriate when map features are separated in different layers and you are converting high quality data. If you have multiple layers in a single raster image you may spend more time editing than digitizing.

Interactive Mode

Interactive vectorization (tracing) is far more user controlled. Here, the operator selects a starting point and the vectorization software traces the line until it comes to an intersection. At the intersection, the vectorizor waits for you to provide a direction, and then it repeats the process. While this method is often only marginally faster than manual digitizing, it yields computer generated vectors, which are usually of higher quality than manual ones.

Choose your method of vectorization with care. Your goal is to produce the highest quality vectors in the shortest amount of time. Test and evaluate each method at the outset; it will save you time in the long run. Use batch vectorization for complex contour maps and with single features on Mylar separates. Use interactive vectorization when you have multiple features or low quality source maps. If you plan to consistently capture the same type of data, you may only require one vectorization method. However, if your data capture needs are diverse, your data capture methods must be diverse as well.

To choose a vectorization package successfully, keep one term in mind: configurable. Many vectorization packages are designed to treat all lines equally and capture them in the same manner every time. While this makes sense in engineering, where all lines are manufactured and essentially uniform, GIS involves capturing cultural and natural features. Natural features are anything but uniform, and each feature type must be dealt with differently. Glancing at a map reinforces this distinction; roads differ greatly from contours, which differ greatly from streams. You do not want a vectorization package designed for capturing straight lines and regular curves when dealing with irregular map features. You must be able to correctly configure the software to capture streams, contours, roads, and facilities. The software must be flexible.

The best vectorization packages use a predefined parameter set that you can customize for each feature type. When creating parameter sets, you must account for a feature's anomalies. For example, roads tend to have a lot of right angle intersections and straight lines. Therefore, you want a vector package that can clean up right angle intersections well. However, because streams have irregularly curved lines with oblique intersections, you want a vectorization package that can add vertices to accurately capture the trend of the stream bed, but also maintain the quality of intersections. With high quality software you can create a different parameter set for each feature type.

Capturing lines accurately requires that you be aware of the two main problems with vectorization: intersections and generalization. Your post editing will be divided between cleaning up poor intersections and adding or deleting vertices where a curve has been captured inaccurately. Correct vectorizing parameters will provide a balance between line generalization and clean intersections.

Intersections

Intersection problems come in two basic types: X intersections (where lines bisect each other evenly) and Y intersections (where lines meet at oblique angles). Both are problematic because the vectorization software is attempting to determine the precise center of a set of pixels comprising the intersection, but the precise center of the raster is not the logical termination point of the intersection.

Generalization

Generalization is the second major problem encountered when vectorizing. The correct number of vertices in a vector line is necessary to accurately depict the features you capture. Too many vertices will result in an irregularly shaped line, and too few may cause discrepancies in accuracy. The number of vertices in a vector line is determined by the type of line you want to represent. Capturing contours entails many vertices because contours are natural features that can change directions dramatically, but capturing roads or parcels will require fewer vertices because these are cultural features with predictable direction changes. To make such adjustments correctly, you can control the amount of generalization, depending on the feature type captured.

Most vectorization packages control generalization using a simple step algorithm that places vertices at regular intervals regardless of the trend of the line. Step generalization can be difficult to control and requires testing in order to achieve the correct step. Some packages use other algorithms that alter the number of vertices according to a line's trend. This process, which results in fewer vertices in straight areas and more vertices in curved areas, is the best type of generalization for GIS. Such algorithms can adjust to roads, for example, which tend to be straight in cities and towns, but follow natural terrain in rural areas. Because you are likely to encounter both types of lines in a given map, the amount of line generalization must vary accordingly.

Intersections versus Generalization

In many cases, intersections and line generalization are inversely related. In other words, your vectorization software may handle curved data with no intersections quite well or straight lines with many intersections. Unfortunately, such packages can rarely handle both problems at the same time because intersection cleanup often relies on straight lines. The more curvature you allow for in a vector line, the lower the quality of intersections.

When shopping for a vectorization software package, send sample data sets to the vendor and have them vectorize the data for you to determine how well the software performs. Ensure that the vendor sends you raw vectors that have not been cleaned manually after the conversion because your goal is to determine how much post editing will be required. Next, avoid obtaining sample data from the vendor because they are likely to skew the results. For example, if you intend to capture natural terrain features, you are not interested in how well the software handles parcel data.

Many vendors will allow you to obtain an evaluation license for their products. Take advantage of such opportunities and combine the evaluation period with data testing at the vendor site. Make the vendor demonstrate that she can capture your data correctly, and then obtain the parameters used in order to repeat the process with the evaluation license. If the capture works the way the vendor claims, she will sell software and you will have a vectorization package that meets your needs.

Attributes

The final issue with vectorization is coding attributes. You want attributes to be captured quickly and accurately. To achieve this, you must have a sound approach for coding. While many vectorization packages can help you with this task, be careful because software vendors may emphasize automatic attribution at the expense of vector quality. In most cases—and particularly when using batch mode—you should leave attribution out of the vectorization process. In interactive mode, however, it can be very efficient to code attributes on the fly by selecting a line symbol or CAD layer to store lines as you trace. Because you control what you trace, you can change layers at will. Most interactive vectorization software packages have this functionality, so coding as you go is not a problem. While you will not be able to code every file in this manner, it can help you with selected layers and save time.

Vectorization Tricks of the Trade

Regardless of the method you use to vectorize, several simple techniques can make capturing data easier.

1. Separate map features. If possible, try to obtain source data with only one feature per scan. This will make vectorization and attribute assignment much easier, and it also allows you to place data into different layers for easy access.

2. Do not compromise between composited features. Because obtaining feature separates for a map is frequently impossible, you will likely find several features on a single scan. If you are using interactive vectorization this is not problematic, but in batch vectorization you may find the quality of one line type suffers when you are capturing more than one. Furthermore, if you try to compromise scanning and vectorization parameters to obtain the best from each feature type, you might degrade every type, making them all useless. Instead,

perform separate scans for each feature in the document, tailoring your vectorization parameters each time.

3. Prep confusing symbols. Many line symbols, such as dashed lines, will not vectorize well no matter what you try. You can save time and frustration if you prepare complex symbology in advance so that it will scan and vectorize properly.
4. Fill in or mask screened polygons. Polygon data on source maps can be extremely problematic. Vectorization does not know how to handle certain screens, and some scanners cannot transfer data fast enough to accommodate large filled areas. If your screened polygons are small, use a water based marker to fill them. With large polygons (such as the ocean area of a coastal map), use paper to mask the entire polygon.
5. Raster process screened lines. If linework itself is screened, you might want to draft through the lines using a marker, but this can be time-consuming. In such situations, consider using a raster processing package with a line smoothing function to fill in raster lines automatically.
6. Remember the basics. The old ways are often best. Some features may require drafting, especially on complex maps. On-screen digitizing is a viable solution if vectorization fails or your source is a paper lithograph. If every avenue fails, you may need to return to the digitizing board.

Color Scanning

Until recently, color scanning was too expensive for most GIS applications. Color scanners are currently fairly cost effective, and as a result they are quickly becoming commonplace. A color representation of a map can be very helpful in capturing data in context with other map data. Color allows you to see how features relate to each other, which can be efficient for building data and assigning attributes in special situations.

Color Scanner

A color scanner produces an image that has three color values for each pixel. There are three basic types of color data: true color, pseudo color, and normalized color.

True color images represent up to 16,777,216 possible colors. These colors are contained in three bands—red, green, and blue. Each band is similar to a grayscale image because it divides each color into 256 distinct values (i.e., the red band contains 256 possible values for red). When the three bands are combined, each pixel can display up to 16,777,216 colors (256 x 256 x 256)—far more colors than the human eye can distinguish. While true color provides the most accurate representation of an image, true color files are very large and can be difficult to manage, even when compressed.

Because most maps are not printed with 16 million colors, true color is usually overkill unless computer analysis is involved. Most scanning uses pseudo color, which produces images that, like grayscale images, are eight bit or reduce 16 million colors to 256. This process degrades images to some extent, but the effect is hardly noticeable. Pseudo color image files are about one-third the size of true color image files. Pseudo color images also tag all three RGB values to each pixel, eliminating the need for three bands.

Pseudo color provides the best compromise between image quality and file size, but true color and pseudo color can be displayed only on the screen. In other words, you can view them, but you cannot extract data accurately with automatic vectorization because the images contain so many colors. To eliminate this problem, you must normalize the image, or reduce each feature to a single color.

Normalization

To understand how normalization works, you must understand in greater detail how colors on a map are represented digitally. Color and grayscale scanners work by seeing light and evaluating its intensity. Color scanning differs because the scanner separates light intensities into red, green, and blue values. Color CCD technology can perform this with a single pass of the scanner because CCD elements are keyed to single colors and three elements are required to determine the value of a single pixel.

Because the scanner is seeing levels of light, each pixel has a unique value—or, more precisely, three unique values corresponding to the three colors. When pixels of differing but similar values are placed next to each other, they appear to be a single color from a distance. In other words, a red line on a map can consist of several values of red. The value seen by the scanner is determined by the intensity of color reflected back to the CCD element, which is in turn determined by the colors surrounding the pixel being viewed. Therefore, the edge of a red line has a different value against a white background than against a green one. The result is an image with several values that constitute a "single color" feature. In order to vectorize a color image effectively, the number of values in a feature must be reduced to one only. The process for accomplishing this is known as color extraction.

Color Extraction

Color extraction can be problematic because many of the values that make up red also make up blue, magenta, and other colors. Therefore, if you extract only red lines comprising highways you will also pick up pieces of streams and populated areas, which show up as clutter on your output image. There are two basic approaches for avoiding this problem: color training and color reduction.

Color training involves an operator viewing the image and discretely selecting color pixels representing a given feature. Neural networks are often created so that the computer can automatically decide which colors in other areas of the map are consistent with the operator's choices. While effective, this process can be time-consuming. Color training software is most effective when you have few features to extract for a large number of similar maps.

The other approach, color reduction, is faster and easier but does not produce results of the same quality as color training. Color reduction, a feature available with most scanner software, involves reducing all colors from 24 bit to eight bit, and then to four bit, or 16 final colors. The best approach for color reduction is to let the computer automatically reduce the colors to 64, and then manually group similar colors together by evaluating them yourself. The colors should be grouped until you have only 16 colors in the image. Many scanners will allow you to make a four bit color palette and will automatically produce the same 16 colors on future maps that you scan. However, you must create different palettes for maps containing different colors.

Optical Character Recognition

Good News About OCR

Optical character recognition (OCR) is software that automatically recognizes raster features and translates them into intelligent vector, tabular, or document data. OCR is currently being used in document management applications to allow users to scan text and automatically translate it into word processor documents. OCR works by matching font patterns to similar raster patterns, or vectorizing the raster and calculating the differences between the new vector and the existing text font. Some OCR packages use very sophisticated artificial intelligence, while others use simple pattern matching. The best OCR packages allow the user to customize character, feature, and symbol recognition and assign attributes to a database or place layers in a CAD drawing. There are a number of vectorization packages on the market that contain OCR functionality. Ask your software vendor to recommend an effective OCR package.

Bad News About OCR

Unfortunately, OCR does not work very well on most GIS maps. Most OCR packages available are font based, and maps rarely use standard fonts. Even the packages that recognize hand written text will only recognize certain types of text. Because map data are so diverse, you are unlikely to find an OCR package that will work with your particular map set. However, all is not lost. If you are converting engineering drawings with standard Leroyed text, your chances of finding an OCR package that will correctly recognize most text are good. But if your maps are non-standard, you will probably need to manually input text.

Even when OCR does work on map text, you may not like the results. If you have a text separate with no other features and your OCR package recognizes the font, you can make good use of OCR. Unfortunately, for most users, such separates are simply unavailable. Because OCR may recognize map features other than text as text, you may spend more time cleaning improperly recognized features than you might if you digitize your text manually.

Of course, OCR technology is always improving, and it has made great leaps in engineering and facilities applications in recent years. At the time of this writing, OCR has yet to be successfully applied to large mapping projects, but eventually you can expect to see success in the GIS arena as well. The key is patient evaluation. If you are intent upon applying OCR in your application you must research the technology and stay informed regarding advances. Do not rely on software developers to provide you with complete information unless you ask specific questions and perform your own tests. Purchasing an OCR package off the shelf could prove costly if you have not fully tested its capabilities.

Even when the technology does become successful in GIS, experience in engineering and facilities management applications has shown that OCR will probably not be 100 percent effective. Customization and post-processing will likely remain a part of the process. If you are currently successfully using OCR in an application, you can maximize your success rate by obtaining high quality scans and limiting the amount of features on source maps by masking unwanted data. In this area, you may rely heavily upon the software developer's technical support.

Data Storage

Scanning applications requires substantial disk space. Compared to vector data, raster images are enormous. To determine how much disk space you will require, you must assess the amount of scanning you plan to do and multiply that figure by the size of your output images. Combine the rate at which you intend to scan with the length of time your data must remain on line, and you will have a fairly accurate estimate of your minimum disk requirements. The calculation becomes more complex when you include performing raster processing and maintaining duplicate images.

To estimate disk usage, you must understand how large raster images can become. The safest factoring method is to assume that all images will remain uncompressed. While the size of an uncompressed image can be calculated precisely, compressed image sizes cannot. To calculate the size of an uncompressed image, multiply the physical dimensions of your document by the square of the resolution. That is, a 24 x 36" map scanned into a 200 dpi pseudo color image will be 24 x 36 x 200^2, or 34,560,000 bytes (roughly 34.5 Mb). If you are scanning for true color, triple the value, which makes your output about 104 Mb. In other words, you will need more than 100 Mb of free disk space to scan a single image. A 2 Gb hard disk will only accommodate 20 such images.

Generally, you will not be scanning uncompressed true color images only, so the outlook may not be quite so bleak. Bi-level images are comparatively small. You can use standard compression to reduce the file size of a bi-level image to about 10 percent of its original size. Compression can also reduce pseudo color and grayscale images to about 30 percent of their original size. Be very careful about the compression method you use, however. GIF and JPEG compression are not considered non-invasive, and will reduce the quality of your image. If you are scanning for data conversion limit yourself to compression algorithms used in standard Packbits compression or Group 4 TIFF compression. These are industry standards and are supported by nearly all scanning interface software packages and most imaging software packages.

Raster processing can also reduce the size of your image by deleting data or merging similar values. It is easier for compression algorithms to reduce data if there are fewer image values to compress. The drawback of raster processing, however, is that many packages require three times the size of the original image in available disk space to load the image. This is because many imaging packages use redundant data to save processing time and eliminate data loss should the package unexpectedly terminate.

Making the Most of What You Have

Once you have realized how much disk space scanning requires, you will soon discover that the amount you have is never enough. You must maximize the disk space you do have by keeping data on line only as long as necessary. Consider your scanning hard disk as a data delivery system, not a data storage system. All scanned data must be transient. Transfer the data to the application storage system and delete it from your scanning system as quickly as possible. You will not want to maintain your raster data on line at all times even in your application storage system. Archive what you do not need immediately and load onto your system only what you will be using.

You can further maximize disk space by creatively storing, processing, and transferring data. Group jobs into batches so that several images reside on a single disk at a time. While you are scanning you will not be able to process data because the scanner will take up most of your memory and much of your processor. Maintain different machines with their own disks for scanning, raster processing, and vectorization. Scan five to 10 maps and immediately transfer them over to the raster processing machine. While the scanner is working on the next batch of maps, begin processing the first batch. When the raster processing is complete, move the images over to the vectorization machine and move the second batch of scanned images to the raster processing machine. This assembly line approach will maximize disk usage and efficiency.

Data Delivery

Once your data are scanned and vectorized, you must load them into the application. The easiest way to accomplish this is from hard disk to hard disk such as through network downloads. Data can be transferred over large distances through an FTP server. If this method is inappropriate or unavailable, you must find another medium.

While floppy disks may come to mind first for data transfer, they are useless with large raster data files. However, some vector files can be transported this way.

Digital archive tapes (DAT) can store large amounts of data and are fairly reliable. They are also fairly inexpensive. Consequently, they are generally considered an acceptable means of transferring raster data. The major drawback to DAT is system compatibility. A DAT must be created on the same platform it will be used on. If you are scanning on a PC and your client is on UNIX, you may have difficulty transferring data via DAT. They are also easily damaged and require a long time to create and download.

Clearly the most versatile method for transferring data is the write once read many (WORM) compact disc writer. CDs are virtually universal. If you write data to a CD they can be read from any CD-ROM drive on any platform. CDs can also store large amounts of data and are perfect for images. Because they are write protected, there is little chance of data loss. The only drawback to WORM drives is cost. CD writers tend to be far more expensive than tape drives.

Summary

Scanning is still new to GIS, but with proper research and foresight, you can put it to effective use in virtually every GIS application. It is not nearly as complex or difficult as it may seem. Like any form of cartographic data capture, scanning is an art form, or at the very least a learned skill. If you are creative and motivated you can develop new and inventive ways of using scanning to support your GIS applications. It is not always the most appropriate method, and you must apply the same wisdom in scanning that you would in other methods of data capture. Understanding basic data capture techniques and a firm grasp of cartography and imaging are required. The key to effective scanning is to understand map data and know how you want to use them, so that you can make appropriate decisions to create accurate results.

Acknowledgments

Thanks to Daryl Smith and Al Derosier of the Environmental Systems Research Institute, Inc.; Jay Magenheim of Ideal Scanners and Systems (Rockville, Maryland); and John Cortese and Dean Dietrich of Scan-Graphics, Inc. (Broomall, Pennsylvania).

Keyboard Entry of Attribute Data

Edward Kura

The conversion of attribute data, while important, can be overlooked in the data conversion process. Spatial data, the presentation component of your GIS, are often considered more important because they are used to create maps and displays. Attribute data are displayed using graphics because they are often easier for most map users to understand. However, information extracted from your GIS is only as reliable as the attribute data within it. Using proper procedures to enter attribute data can ensure that your GIS database will contain reliable information.

The digitizing technician who captures spatial data may perform keyboard entry of attribute data at the same time, or attribute data may be handled by a separate data entry operator. In conversions involving high volumes of data, the latter method is generally preferred. While digitizing is performed on a high end personal computer or workstation equipped for computer aided design (CAD), attribute data entry can be performed on a desktop PC suitable for running a spreadsheet or database program. Digitizing technicians generally require extensive CAD training, while data entry operators can be trained in a few hours. Student interns or co-ops can be a cost-effective staffing resource for keyboard data entry. The combination of an inexpensive PC and lower staffing costs make the low tech data entry option an attractive choice for many projects.

This chapter discusses two options for keyboard entry of attribute data: single key and double key. Both options use identical hardware and software, and can be performed by the same staff. While double key entry may imply that it takes twice as long. As such, it may in fact be the faster solution.

One element common to single key and double key entry is the use of a primary key, or unique identifier for each digitized feature, which is entered as spatial data are digitized. The primary key, also entered during both keyboard entry methods, is used later in the conversion process to associate features and their appropriate attribute sets.

The primary key links features (or graphics) to their attributes.

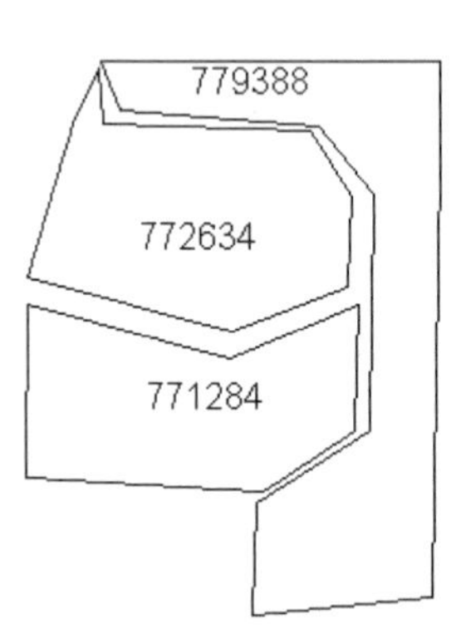

Table of Attributes

Primary Key	Soil Type
771284	A
772634	A
779388	B

Another concept used in both approaches is called logical checking of attributes. In any database design, relationships can be defined between attributes. For example, in a water distribution system GIS, the water main type (for transmission or distribution) might be defined in terms of the pipe size (less than 16" for distribution, and 16" or greater for transmission). The set of possible values for each attribute can also be restricted to values contained in a look-up table, if a finite set of values for the attribute is predefined. A look-up table is simply a list of acceptable values for an attribute. For example, the look-up table for a water pipe material attribute might be restricted to cast iron, ductile iron, PVC, and concrete; no other values would be accepted by the GIS. Logical checking can be performed during single and double key approaches.

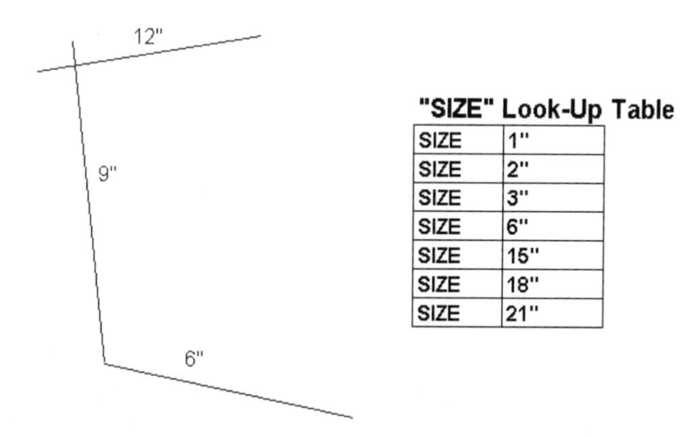

"SIZE" Look-Up Table

SIZE	1"
SIZE	2"
SIZE	3"
SIZE	6"
SIZE	15"
SIZE	18"
SIZE	21"

A look-up table of legal attribute values for size.

The following sections describe the advantages and disadvantages of each approach, the projects for which a given method is preferable, data preparation requirements for each method, data validation considerations, and quality control approaches common to each method. For more information on quality control see Chapter 16, "Quality Control/Quality Assurance."

Single Key Data Entry

Under single key data entry, attribute data are keyed in once. The data are then checked twice for quality control: once using a software routine appropriate for the task, and once by a second data conversion technician. The single key approach is preferable when PC time is limited or when you want to curb personnel costs. It is also preferred in cases of relatively few attributes to be entered per feature. The following sections provide a detailed discussion of the single key data entry process.

Data Preparation Requirements

Data conversion can collect data from many different kinds of paper sources. Source complexity and legibility can vary widely, even for a single data type within a project. Data must be interpreted or "scrubbed" before they can be entered into a GIS. Using single key data entry, scrubbing is performed by the data entry operator as data are entered. The operator evaluates the information presented on the source document and enters it into appropriate data fields. Operators must possess an in-depth knowledge of the source documents and acceptable values for each attribute. Illegible, questionable, or missing data can also slow down the data entry process. Scrubbing during data entry is appropriate for smaller data conversion projects.

For large data conversion projects, you may prefer to make scrubbing a separate operation. In this case, scrubbing involves color coding attributes that will be entered later by a data entry technician. Color coding makes it easier for the technician to select attributes, thereby improving the speed and accuracy of the process. The scrubbing technician also performs all research required to ensure that attribute sets are complete. Research should include locating missing data, resolving questions raised when data attributes conflict, and solving legibility problems encountered on source documents. The end result of the scrubbing process is a source document ready for use by a data entry technician.

When possible, the scrubbing operation should involve people who are normally caretakers of the paper sources being converted. They are familiar with the data, and can easily answer questions and fill in gaps. In most data conversion projects, people are an important source of data because they carry informal knowledge of paper sources that is rarely in written form.

Hardware/Software Requirements

A standard office desktop PC, designed for word processing and spreadsheet use can support single key data entry. The data can be keyed into a spreadsheet or database. Tables in word processing software are not recommended because they are generally less robust than spreadsheet or database programs. After data entry is complete, the data must be transferred to the computer storing the GIS database. Ideally, the PCs being used to create your GIS are networked. Keep in mind that backing up data is important at all stages of the conversion process, not just when the database is complete.

NOTE: *Networks also simplify backup procedures for all data generated as a GIS database is created.*

Some programming may be necessary to create data entry screens in your database or spreadsheet software. The level of sophistication of such screens should be consistent with the skills of the data entry operators. Operators with limited PC or data entry experience will require simple screens which may nonetheless hide sophisticated programming that automates as many functions as possible.

Simple data entry environments can be created using macros in spreadsheets, while more sophisticated environments may require you to develop code in an object oriented programming language. Pick lists or drop-down lists in the graphical user interface can be used to restrict data input to a set of legal values. Whether you use macros or more complex programming, the interface becomes easier to use. With ease of use, accuracy and speed of data entry will improve, driving down the costs of data conversion.

Data Validation Considerations

Key entered data should be validated during the data entry process to ensure that they conform to the rules and standards established in your GIS database design. This validation includes the logical checking of attributes described previously.

Data validation and logical checking can be performed as the data are entered or afterward. On-line validation, while it may slow down the process, alerts the operator as soon as it detects an illegal attribute value and requires a correct entry before allowing more data entery. In some systems, only an override key combination can be entered to force the software to accept data that does not conform to a predefined standard. This is particularly useful when the technician is locating the attribute data from sources, which is common in single key data entry. Data validation after data entry is covered in the section on double key data validation.

Data validation programming should be part of project planning for the software platform used in data entry. As with the design of the data entry environment, more sophisticated validation routines improve the accuracy of attribute data and may eliminate future problems regarding how attribute and digitized graphics data are joined, or linked.

Data Quality Control

The level of quality control you perform should be established as early as possible in the project. Quality control can have a major impact on the integrity of your GIS database, the level of effort required, and the overall project budget. The pilot phase of the conversion is an appropriate time to define the quality control standards that will be enforced for the balance of the project (see Chapter 4, "Project Planning and Management").

You can adjust the level of quality control set for your project in response to several factors, including source data quality and staff performance. However, these changes should not be taken lightly or made frequently. You should strive for an even quality standard throughout the database, and this goal may be difficult to achieve if quality control standards are altered. Furthermore, while you can loosen standards at any time (although it will impact the quality of your database), tightening standards will require reviewing all data converted up to that point to ensure that they meet the new standard. Drafting tighter standards could negatively impact on your project schedule and budget, especially if large amounts of data have already been converted.

Quality control checking of entered data involves logical checking and manual checking. Logical checks are performed as data are entered and errors are corrected before the process continues. Exceptions to predefined standards should be logged as they are entered for future reference because subsequent quality control procedures or quality assurance checks may flag the values as errors. A log explaining the exception may save considerable time in research later.

Manual checks are critical because they locate errors that were undetected in the logical checking phase. In many data conversion projects, this latter phase is the only time when attribute data are checked against original sources for accuracy. To help you budget for this essential procedure, estimate that manual quality control will take 50 percent as long as the actual data entry.

Manual checking and data entry should be performed by different individuals otherwise you run the risk of having the same person interpret data incorrectly twice—once during data entry, and again when performing quality control checks. While it is possible that an individual may not repeat a mistake, you can significantly reduce the possibility if you separate the tasks.

Two methods are typically used in manual quality control. In the first method, the quality control technician compares a printed report of entered data with source documents. In the second method, known as line quality control, the technician reviews entered attributes on the computer screen and compares them with source documents.

To perform checks using printed reports, the reports must be generated after all data in a particular work unit are entered. Defined as one or more sources with a common element (such as adjacency), a unit of work should be small enough that it requires eight hours or less to review. Procedures for generating printed reports and correcting errors should be established when the data entry software is written and other procedures are established.

Performing quality control using printed reports involves checking information in the printed report against source documents. All attribute information must be verified. Errors should be clearly noted on the report, along with correct values. Where possible, database corrections should be made by the same technician who completed the original data entry. This procedure provides technicians with the opportunity to review their errors and helps them learn from their mistakes. After the corrections have been entered, the quality control process is repeated until no further errors are found or the standard is met.

On-line quality control does not require printed data reports. A technician views each database record on the screen, compares it with source documents, and immediately corrects errors. Automated and logical checks should be performed after on-line quality control is complete, particularly to find errors inadvertently introduced during the quality control process. Because on-line quality control does not involve a separate procedure for checking corrections (other than repeating the process again), technicians must take extra care during on-line quality control.

Double Key Data Entry

The second method of data attribute entry, double key, requires that all attributes be entered twice by two different operators. The two data files are then compared using a software routine, and the differences between them are flagged for action by a third quality control technician. Once discrepancies are resolved, a final copy of the data is written into a new file and data validation processes are performed. After all procedures are complete, the data file can be loaded into the the GIS database. This method relies on the premise that two operators will not repeat the same data entry errors.

The double key approach is preferred when large volumes of attribute data are involved. A data scrubbing process preceding the data entry phase produces forms that data entry operators use to input data. Because operators enter the data directly from these forms, they need not have expert or extensive knowledge of the data or database. In other words, the skill level of individuals performing data entry is not an issue. Data entry costs are reduced because temporary or existing staff can complete data entry. The following sections provide in-depth discussion of the double key data entry process.

Separate data files are compared and merged using software to yield the final output.

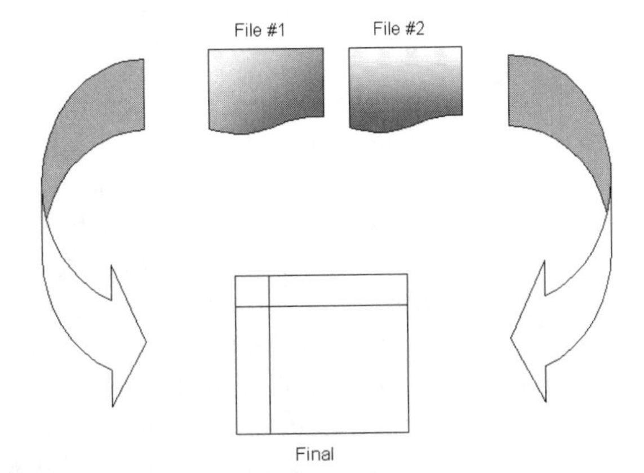

Data Preparation Requirements

Data preparation for double key data entry is far more intensive than single key data entry. In the double key approach, operators work exclusively from data forms, not original source documents, which changes the process considerably. Operators do not require knowledge about sources, the database, or the GIS. They merely must enter information exactly as it appears on the forms. This shifts the responsibility for interpreting source data to technicians performing data scrubbing. Data entry operators input data from forms prepared by the scrubbing technicians and only question the data when the software rejects the values they enter.

During the scrubbing process, attributes are copied from original source documents onto data forms. Prepare a data form for each feature being entered (e.g., water main or valve), including information for each attribute that must be added to your GIS database. All questions and conflicts regard data are resolved during the scrubbing process. Prior to data entry, quality control must be performed on each form to ensure that the data are error-free. This check is even more critical under the double key entry method because the data forms become the only source used by data entry operators.

✓ **TIP:** *Clear, accurate data forms are essential to successful double key data entry.*

The following illustrations show different methods of cataloging data for double key entry. The single feature data form contains attributes for one feature only. The technician selects the correct value for an attribute by circling one from a list of allowable attributes on the form. The multiple feature data form contains a table of columns (attribute data) and rows (feature data), and defines default values for each attribute. You may prefer the multiple feature form for high-speed data entry because operators are not required to turn the page for each feature. However, the single feature form is easier to read. In either case, computer data entry screens should resemble the forms, which in turn should reduce training time and improve accuracy.

SAMPLE -- SINGLE FEATURE DATA FORM

STORMWATER INFRASTRUCTURE INVENTORY

MAP NO. ___________

STORM CULVERT

Completed by:________________Date: ___________

Checked by:________________Date: ___________

ID Number

Upstream ID Number	Default = Unknown	____________________________
Downstream ID Number	Default = Unknown	____________________________
Barrel Depth	Default = 36"	____________ Inches
Barrel Width	Default = 36"	____________ Inches
Number of Barrels	Default = 1	____________
Barrel Material	Default = Concrete	Conc. CI CMP RCP PVC Unknown
Inlet Material	Default = Grass Channel	Grass Concrete Stone/Block
Outlet Material	Default = Grass Channel	Grass Concrete Stone/Block
Headwall Shape	Default = Flat	None "U" "L" Flat Flared
Owner ID	Default = City	City Personal Property
Structural Integrity	Default = Unknown	Sound Partial Collapse Full Collapse Unknown
Obstructions	Default = Unknown	Clear <50% Blocked >50% Blocked 100% Blocked Unknown
Obstruction Type	Default = Unknown	None Woody Debris Fine Sediment Trash Coarse Sediment Other Unknown
Erosion - Upstream Failure	Default = Unknown	None Moderate Extreme Potential
Erosion - Downstream	Default = Unknown	None Moderate Extreme Potential Failure
Concrete Class	Default = Unknown	C75 Concrete C76 Concrete Unknown
Upstream Elevation	Default = 999.9	_______________ Feet
Downstream Elevation	Default = 999.9	_______________ Feet
Slope	Default = 999.9	_______________ %

Notes: (oil & grease, unusual odors, colors, turbidity, dumping, or evidence of flooding)____

____

____

SAMPLE -- MULTIPLE FEATURE DATA FORM

SEWER INFRASTRUCTURE INVENTORY

Facility: SEWER MAIN

Page 1 of ________

Map No.: ________ **Scrubbing Tech. Initials:** ______________ **Scrub Date:** ____________

Quality Control Tech Initials: __________ **QC Date:** _________

ID #	Length D=0	Type D=GR		Owner D=CI				Size D=0.0	Material D=U				Year Installed D=0	Function D=CO			Status D=AC		Upstream Invert Elev. D=999.9	Downstream Invert Elev. D=999.9
		GRAVITY	FORCE	PRIVATE	STATE	COUNTY	CITY		CONC	CLAY	CI	U		COMBINED	SANITARY	STORM	AB	AC		
		GR	F	PR	ST	CO	CI		CO	CL	CI	U		CO	SA	ST	AB	AC		
		GR	F	PR	ST	CO	CI		CO	CL	CI	U		CO	SA	ST	AB	AC		
		GR	F	PR	ST	CO	CI		CO	CL	CI	U		CO	SA	ST	AB	AC		
		GR	F	PR	ST	CO	CI		CO	CL	CI	U		CO	SA	ST	AB	AC		
		GR	F	PR	ST	CO	CI		CO	CL	CI	U		CO	SA	ST	AB	AC		
		GR	F	PR	ST	CO	CI		CO	CL	CI	U		CO	SA	ST	AB	AC		
		GR	F	PR	ST	CO	CI		CO	CL	CI	U		CO	SA	ST	AB	AC		
		GR	F	PR	ST	CO	CI		CO	CL	CI	U		CO	SA	ST	AB	AC		
		GR	F	PR	ST	CO	CI		CO	CL	CI	U		CO	SA	ST	AB	AC		
		GR	F	PR	ST	CO	CI		CO	CL	CI	U		CO	SA	ST	AB	AC		
		GR	F	PR	ST	CO	CI		CO	CL	CI	U		CO	SA	ST	AB	AC		
		GR	F	PR	ST	CO	CI		CO	CL	CI	U		CO	SA	ST	AB	AC		
		GR	F	PR	ST	CO	CI		CO	CL	CI	U		CO	SA	ST	AB	AC		
		GR	F	PR	ST	CO	CI		CO	CL	CI	U		CO	SA	ST	AB	AC		
		GR	F	PR	ST	CO	CI		CO	CL	CI	U		CO	SA	ST	AB	AC		
		GR	F	PR	ST	CO	CI		CO	CL	CI	U		CO	SA	ST	AB	AC		
		GR	F	PR	ST	CO	CI		CO	CL	CI	U		CO	SA	ST	AB	AC		
		GR	F	PR	ST	CO	CI		CO	CL	CI	U		CO	SA	ST	AB	AC		
		GR	F	PR	ST	CO	CI		CO	CL	CI	U		CO	SA	ST	AB	AC		
		GR	F	PR	ST	CO	CI		CO	CL	CI	U		CO	SA	ST	AB	AC		
		GR	F	PR	ST	CO	CI		CO	CL	CI	U		CO	SA	ST	AB	AC		

Hardware/Software Requirements

A standard office desktop PC designed for typical word processing and spreadsheets will accommodate double key data entry. Programming should be developed to support the functions listed below.

- Data entry screens
- Data validation routines
- Modules to compare two sets of keyed data, and produce reports comparing the two
- On-line corrections for both data entry files
- Creating a final, corrected version of the data file

The software should also allow for network storage of all data entry files, if available, in order to include data file backup in your normal backup process. If you are not on a network, schedule frequent backups for all computers storing data entry files.

Once data entry is complete under the double key method, the data must be transferred via floppy disk, the network, or other medium, to the computer storing your database. As with single key data entry, the process is easier if all PCs being used to create the GIS are networked.

The same rules regarding computer environments for data entry under single key apply under double key. Simple environments for inputting data can be created using macros in spreadsheets or a modern, object-oriented programming language for more sophisticated environments. Creating such an environment accelerates data input and improves accuracy. See the "Hardware/Software Requirements" in the section in this chapter on single key entry for more information.

Data Validation Considerations

Data entered under the double key method should be validated only after both sets of data are entered and compared, and a third, final data file has been created. Executing the validation process on two separate data files is redundant and does not improve data quality. As with the single key process, validation includes the logical checking of attributes described in the first part of this chapter.

Data validation programming for double and single key data entry have the same qualities: more sophisticated routines improve the accuracy of attribute data and can eliminate future problems regarding how attribute and digitized graphics data are joined or linked.

You must define acceptable values for each attribute as well as relationships among attributes as part of the database design process. The criteria you identify will be used in the logical checking phase to ensure that attributes comply with predefined sets of acceptable values. Where no predefined set of values has been established, you may be able to define other parameters, such as an allowable range of numbers, or only positive numbers between 5 and 99. In most database designs, the only attributes for which logical checks are ineffective are free form comment fields, which typically hold miscellaneous information that cannot fit more formal attribute definitions. For this field, you may want to run a spell checker rather than logical checking.

Data Quality Control

Quality control for double key data entry is performed in two phases. Once the data are scrubbed, manual quality control—the more important phase—occurs. This involves checking all attributes entered onto the data forms against source documents. Remember to assign individuals for this process different from than those who scrubbed the data. At this time, you should also resolve all other questions, such as illegible information or conflicting attributes.

The second phase of quality control takes two steps. The first step, performed by software, compares the two data files. The software flags errors, which are corrected by an operator. Repeat this cycle until no errors are detected. The software then writes a new, third version of the corrected data file. At this point, in the second step, the software again performs logical checks on the corrected data file. Attributes that do not pass the logical check must be researched and corrected. Repeat this process after each round of corrections, until you are confident the data are accurate. Maintain a log of all exceptions and special cases, if applicable. You may need it later to explain why an attribute does not conform to established standards.

You should expect to modify quality control procedures during and after the pilot project. As you work with the data and the database design, more logical relationships will become apparent. Incorporate as many of these relationships as possible, within budget constraints, into applicable software routines. Logical checks are inexpensive to program, and because they are performed automatically do not add significantly to the project's cost. When logical checks are very complex and require substantial computer time, you may want to run them on a separate machine or at night.

NOTE: *There is a direct relationship between the number of logical checks you perform and the quality of the final database.*

You may choose to alter quality control procedures to improve efficiency or quality. The level of quality control can be adjusted during the conversion effort in response to several factors, including source data quality and staff performance. However, these changes should not be taken lightly or made frequently. You should aim for an even quality standard throughout the entire database, which may be difficult to achieve if standards are adjusted. Furthermore, while you can loosen standards at any time (although it will impact the quality of your database), tightening standards will require reviewing all data converted up to that point to ensure that they meet the new standard. Drafting tighter standards could negatively impact on your project schedule and budget, especially if large amounts of data have already been converted.

Spatial Data Transfer Standards

Phyllis Altheide

Standards for GIS are becoming increasingly important in the rapidly expanding geospatial industry. In the context of data conversion, widely deployed standards are critical to increase data reuse and sharing, and to decrease the effort data conversion can require. One type of standard important to the industry is a spatial data transfer standard, the goal of which is to automate data exchange among different hardware and software systems without losing information.

Such standards address spatial data, or data referenced to the Earth's surface (also known as geospatial or georeferenced data). The transfer of spatial data is defined as the movement of data from one software/hardware system to another. The transfer might be from a data producer to a consumer, or to (or from) a company that adds value or enhances the data. The transfer may be across time (into an archive) or "blind" (without knowing the recipient, such as on the Internet).

This chapter discusses the general characteristics of a spatial data transfer standard, impact on data consumers, practical aspects of importing data, and troubleshooting of common problems. Throughout the chapter, the Spatial Data Transfer Standard (SDTS) FIPS PUB 173 is used as an example.

NOTE: *SDTS FIPS PUB 173-1 is used by U.S. federal government agencies for local or regional data sets. It is endorsed by the Federal Geographic Data Committee as a transfer standard to support a geospatial framework.*

General Characteristics

No Information Loss

A fundamental goal of any data transfer standard is to accomplish the transfer without losing information. For spatial data, the information is composed of far more than just spatial objects and descriptive attributes. It also includes a description of the processing steps, data extraction guidelines used, reports of positional accuracy (desired and actual), a statement of the data's temporal characteristics, and other similar items. The geospatial industry refers to such companion data as *metadata*, which can be understood as data about data.

A transfer standard defines a mechanism whose goal is zero information loss, but there are no guarantees. The data producer must provide complete information—a subjective responsibility that can vary depending on how the data will be used. For example, assume that an end user who plans to supplement a data set with additional information to complete the coverage of a given study area. The user would need to know the data extraction rules so that the coverage is consistent over the entire study area. A variation on this occurs when end users wish to panel adjacent areas together into a seamless piece. These users would also require knowledge of extraction rules and temporal variations.

A standard will often require the inclusion of certain metadata, regardless of the end user application. For example, the SDTS requires that the data set denote a ground coordinate system, so that a user can properly interpret the spatial coordinates of the transferred data. A standard provides a mechanism to accommodate additional information that a data producer or publisher wants to include in a data set.

Self-contained Transfers

A data set created under a data transfer standard should "speak for itself." It should contain all the information you need to assess the data and use them, if you determine they are appropriate for your application. Data sets should also contain information so that importing in specific softwares automatically establishes the data set in a target environment. A data set with these characteristics is "self-contained."

A typical transfer would include spatial objects (e.g., lines, points, areas, and other features). A data producer uses attributes to describe these spatial objects as real world features. For example, a line might be attributed as a stream with the name "Miller Creek" and the status "intermittent." The shape of the line is encoded as a sequence of XY coordinates. In a well-defined system, the parameters to transform these internal file coordinates to ground coordinates are typically encoded in a standard manner. Decoding software can then read these parameters and apply them if necessary to transform the data to another reference system.

A self-contained transfer should not require data users to seek additional information about data contents. To some extent, this self-contained property is a characteristic of all transfer standards. Some standards consider lineage or processing history important and include it, while others are only concerned with describing the end result. The SDTS requires a minimal amount of metadata. However, it provides both a standardized method of encoding other metadata and a mechanism for specifying additional metadata.

The SDTS requires the following metadata.

- Transfer standard's version and profile.
- Ground reference system.

- Data dictionary entry for every non-standard feature type and attribute.
- Data quality report on lineage, positional accuracy, attribute accuracy, completeness, and logical consistency.

The SDTS provides a standard, defined, yet flexible structure to encode the optional metadata listed below.

- Security information.
- Graphical representation information (e.g., symbols, colors, fill patterns, and so forth).
- Directory to all files contained in the transfer.
- Geographical footprint of the data set.
- Other related information.

Finally, if these standard defined items are insufficient, the SDTS permits user-defined attributes and provides places to locate comments throughout a transfer.

Not Intended for Processing

The file structure and format of a data transfer standard are designed to transfer data in a system-independent manner. The structure does not readily support direct processing, other than through very simple viewing and browsing applications.

The structures used in a standard are not intended to replace the internal structures of a GIS. The reason that a transfer standard format is not optimum for processing is because efficient data organization (database design) depends on the application—no single design works for all applications. Consider, for example, the applications of vehicle routing, site evaluation, and a query based on a radius search. For maximum effectiveness, each application requires that you organize data differently. An optimal database design usually takes advantage of various system strengths that a transfer standard cannot address, such as hardware, operating system, and software features. However, a standard must remain generic for portability.

Before you can use a transferred data set in your system, you must first convert it from the transfer file format into a structure compatible with your target system. For this reason, a transferred data set is most often delivered prepackaged. Data providers typically do not implement spatial data transfer standards as interactive client-server data retrieval from a remote spatial database.

Non-proprietary Format

A spatial data transfer standard defines a format that is non-proprietary. No private organization owns it, and you do not have to pay a fee to use it. There are no patents, copyrights, or carefully guarded company secrets to overcome. The standardization process is open to contributions from any individual or entity. A primary objective of the process of changing standards is to ensure backward compatibility with data sets created under earlier versions.

Three Model Levels

Every transfer standard addresses three modeling levels: conceptual, logical, and format. The conceptual level incorporates the philosophy, ideals, and definitions forming the foundation and scope of the standard. The logical level is the collection of abstract data structures and set of information elements that comprise the conceptual level. The format level is the scheme for organizing the data into physical records and bytes.

The SDTS and its profiles contain all modeling levels. The conceptual level defines spatial object types, describes five categories of data quality information, and discusses coordinate reference systems, among other things. The logical level defines the structures of a module, module record, field, and subfield. The format level employs another general purpose data transfer standard, called ISO 8211, rather than inventing a custom file format. The SDTS currently has only one format, but because the logical level was designed independently of the format level, alternative formats may be added in future versions. Because many data transfer standards offer multiple file formats, the manner in which information is encoded into files can be very different even within a single standard.

An SDTS example illustrates the three model levels. At the conceptual level, the SDTS defines a chain spatial object as unidimensional, with topological relationships and a non-intersecting sequence of line segments. This method of describing the properties and characteristics of a feature is called a "chain." At the logical level, you will find a line module with a record representing a chain spatial object. It contains fields for polygons to the left and right of the line, start and end nodes, and a sequence of XY coordinates. At this point, you know the information required to represent the concept "chain," but you do not know what form it takes in a file. The format level defines how to encode a line module record in an ISO 8211 formatted file.

The three levels of a standard in a sample SDTS file. Item A shows the conceptual level; item B, the logical level; and item C, the format level—in this case, an ISO 8211 formatted file.

A.

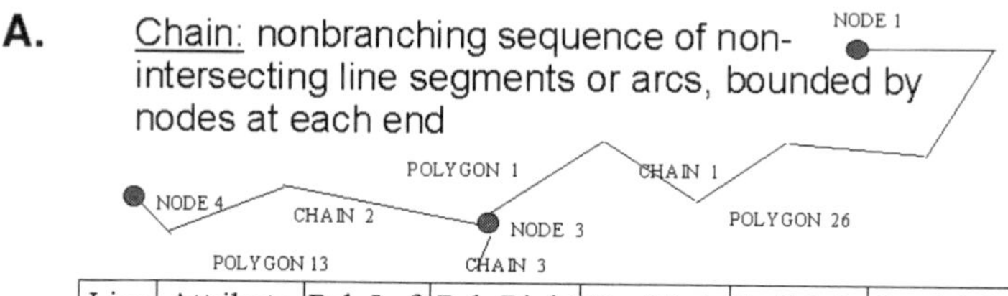

B.

Line	Attribute	PolyLeft	PolyRight	StartNode	EndNode	Spatial Addr
LE01 1	road	PC01 1	PC01 26	NO01 3	NO01 1	123,421 265,526
LE01 2	road	PC01 1	PC01 13	NO01 4	NO01 3	10,425 123,421
LE01 3	.....	.....	....	.....	.....	....

C.

```
004412L 0600106 23040000150000001280 15LINE43043ATID42086PIDL44128PI
DR45172SNID41217ENID39258SADR38297;0000;&HY01LE01;0100;&DDF RECORD IDENTIFIER;
1600;&LINE&MODN!RCID!OBRP&(A(4),I(6),A(2));2600;&ATTRIBUTE ID&*MODN!RCID&(A(4),I(6));
1600;&POLYGON ID LEFT&MODN!RCID&(A(4),I(6));1600;&POLYGON ID RIGHT&MODN!RCID&(A(4),I(6));
1600;&STARTNODE ID&MODN!RCID&(A(4),I(6));1600;&ENDNODE ID&MODN!RCID&(A(4),I(6));
2600;&SPATIAL ADDRESS&*X!Y&((2B(32)));
00162D 00081 220400010700LINE1307PIDL1120PIDR1131SNID1142ENID115
3SADR1764; 1;LE01 1LE;PC01 1; PC01 26;NO01 3; NO01 1;
@@@M@&@(@@&t@&@i; (spatial addresses are binary)
00162D 00081 220400010700LINE1307PIDL1120PIDR1131SNID1142ENID115
3SADR1764; 2;LE01 2LE;PC01 1;PC01 13; NO01 4;NO01 3;
@@@@@&@c@@@M@&@(;
```

Wide Spectrum of Spatial Data Models Supported

Because a spatial data transfer standard is primarily concerned with transferring underlying spatial objects and map attributes, it does not focus on cartographic symbolization or the map image itself. This feature makes spatial data transfer standards flexible enough to support a wide variety of spatial data. Common examples include unstructured vector data, planar graph vector data with full topology, non-planar graph vector data with network topology, raster images, raster grids, and point data. In addition, a standard often supports multiple layers and tiles within a single transfer. Designed to be far broader in scope than a single product specification, standards can accommodate a wide range of spatial data products.

Robust standards are usually implemented through their subsets (profiles); full implementation of a standard is rare. A profile is defined as a clear, comprehensive subset of the standard based on a particular spatial data model. As such, many of the full standard's options are restricted. For example, the SDTS Topological Vector Profile (TVP) features the following requirements.

- Fully topologically structured vector data.
- One of three reference systems: UTM, State Plane Coordinate System, or geographic latitude and longitude.
- Coordinates encoded as 32-bit binary integers.
- Module and file naming conventions.

Feature Catalog

A major barrier preventing the reuse of spatial data is incompatibility among sets of feature types and attributes that will be incorporated into a given data set. To help alleviate this problem, some spatial data standards include a standardized list of feature types and attributes, often called a feature catalog or dictionary. The feature catalog aims to classify features in a generic fashion irrespective of a single application. If permitted by the standard, a data producer can register or define additional feature types within the context of a particular transfer.

For example, the SDTS feature catalog defines the term "watercourse" as a "way in which water may or does flow," but the standard also permits user-defined feature types. The hydrographic feature list used by the U.S. Geological Survey (USGS) includes numerous features that can be classified as a watercourse, including creek, stream, river, canal, aqueduct, wash, and drainage ditch. Depending on your application, you might prefer a detailed hydrology breakdown similar to the USGS list, the general term, or something in between. Spatial data transfer standards either require or highly recommend the use of feature catalogs to ensure more data is reused in the spatial data community.

Files of the Transfer

Presumably, you could determine whether a data set conforms to a standard by inspecting the files, although this is not the most convenient method. Because these files are meant to be read by software, not people, they contain binary and non-printable control characters. However, certain files within the transfer may identify the standard used, and possibly the profile (see the "SDTS Import Example Using ARC/INFO" section in this chapter). Decoding software can read these files and directly determine whether the data set can be decoded based on its standard and/or profile version.

Data transfers involving standards can be organized as single, very large files or a series of smaller files. The single file philosophy packs all information into one file divided into sections. The latter method organizes information in a number of separate files. With either technique, the records (or individual files) contain tags that are detected by the software. Software tools are necessary to manipulate the files because they cannot be directly accessed.

Impact on the Data Consumer

Consider the following scenario: the data for your GIS project have already been collected, you have them in softcopy format, and you are ready to load them into your system. You then load the media, and select Import from the pulldown menu. "Format not recognized" is the reward for your efforts. Six cups of coffee, 10 telephone calls, and four file transfer sessions later, you give up in frustration. There must be a better way, you think to yourself.

As GIS use increases, data sharing problems are growing. GIS usage is no longer limited to cartographers, geographers, and Earth scientists, having become common in the health, agricultural, and law enforcement fields, to name but a few. Spatial data transfer standards are a step in the right direction, but not a cure-all for the numerous barriers to spatial data exchange.

As a data consumer, you may be wondering how spatial data transfer standards will affect you. The following section explores the effects of standards on the GIS software you use and what you must learn. Most data consumers rely on commercial or public domain software written and maintained by others. If you are one of the few data consumers who also writes her own software, a spatial data transfer standard affects you to an even greater extent. Implications for programmers are beyond the scope of this text.

Barriers to Spatial Data Exchange

- Computer media: Do you have the correct hardware for reading the media on which the data are stored (nine-track tape, tape cartridges, floppy disk, CD-ROM, and so forth)?
- Physical format: How are the data written on the media? For example, Windows-compatible PCs and the Apple Macintosh use the same size floppy disk, but their physical formats are different—and incompatible.
- Logical format: What is the data model? What is the record format of the data set? DLG-3 and TIGER/Line are topological vector data models with very different for-

mats. DEMs and DTEDs contain the same type of information, but use different formats. (For more information on these data models, see Chapter 9, "Using Data Sources and Formats.")

- Data content: What type of geographic/cartographic data are contained in the data set? What sources were used to collect the data (e.g., boundaries, hydrography, hypsography, railroads, or timber; high-altitude photography or field surveys)?
- Data completeness: What rules were used for data extraction (e.g., minimum lake size 60 sq yds; minimum tree size 15'; or county and state boundaries, but not municipal boundaries)?
- Data coverage: What geographical area does the data set cover (e.g., city, county, state, country, or continent)?
- Data resolution: What scale are the data intended for? Have the data been generalized?
- Attribution: What scheme is used for attributing the data (e.g., unordered codes or abbreviated feature names)?
- Quality assurance: Were the data subject to verification tests (e.g., positional accuracy, completeness, and accuracy of attribution)?

Target System Must Support Data Set

In order to benefit from a transfer data set, first and foremost the software you plan to use must support or implement the standard. As mentioned previously, a transfer data set must be imported into your software (or target) system before you can access it. Clearly, the most direct transfers occur when the target system supports direct import of the data set.

NOTE: *If the software you use does not support a standard, ask the vendor if an upgrade or add-on software is available.*

If your GIS software does not support a transfer standard, you may need to use a format converter utility that specializes in converting data sets from one format to another. The strategy here is to convert a data set from the source system into an intermediate format that then can be imported directly into your target system. This technique is discussed further in the "GIS Import Process" section below.

What does it mean for GIS software to support a spatial data transfer standard? Implementation can range from very robust to minimal. Very robust implementations are capable of importing (or reading) and exporting (or creating) transfer files. They feature a mechanism through which the user can customize control over the conversion and support the current set of profiles. The software may even be certified as complying with a given standard, if the vendor pursued such designation. On the other end of the spectrum, minimal implementation will support the import of only the data themselves and may import only portions of information from the transfer data set.

As the user, you must determine the extent to which your software implements the transfer standard or profile you require. If the software does not support direct import, the vendor may be able to provide information about third party conversion utilities that operate in conjunction with its software.

Learning the Standard

Telephones, television, fax machines, computers, networks, and other equipment all function in part because of standards. Each requires you to learn something in order to benefit from it, and GIS standards are no different. Learning a standard is a one-time investment that you will benefit from time and again. As a consumer, the amount you must learn is minimal compared to what a GIS software vendor or data provider must absorb. Standards are not light reading, and they are intended for individuals with sophisticated technical skills. What you as the user must know about a standard can often be found in overview articles or similar documentation.

For each GIS standard, you must determine its scope, intended use, reliance on profiles, and the set (and scope, if applicable) of each profile. Knowing this information will help you evaluate potential data sources more quickly. For example, if you want to acquire three-dimensional atmospheric science data, an SDTS raster profile transfer will not suffice because its scope is two-dimensional raster image and grid data. If you find that one standard is intended for archive purposes while another is intended for rapid abbreviated image transmissions, you have an indicator of the level of information found in a transfer data set. A data set bound for an archive will be self-describing, but an abbreviated transfer will require external documentation to complete the information.

Each standard has a vocabulary and set of concepts associated with it. You must learn enough about the conceptual level of the standard to understand the data sources you select for your application. Many standards support multiple data models (e.g., raster image, raster grid, vector, point, and volumetric science data). You would only need to know the profile of the standard that best defines the family of data sources you plan to use. In addition, at the logical level you will encounter information describing how the data are structured. For the SDTS, understanding the terms *transfer*, *module*, *chain*, *GT-polygon*, *2-d manifold*, *foreign identifier*, *reference system*, *data dictionary*, and *data quality report*, among others, will help you understand what it means to receive an SDTS transfer and what information you can expect.

The format level of a standard is typically intended for programmers or GIS implementers, not data consumers. However, you will find it useful to know whether a transfer consists of a single file or many files, file naming conventions, and the recommended usage of directories for managing data files, if applicable. For example, an SDTS transfer consists of many files, including a directory file. The file naming convention required by the SDTS TVP uses the 8.3 convention: eight characters of file name followed by a dot and then a three-character extension. The first four characters of a file name are the same for all files in the transfer, and the last four indicate the SDTS module that the file contains. The file extension is *.ddf.* It is highly recommended that each SDTS transfer be kept in a separate subdirectory.

Spatial Object Terms

Every GIS software vendor has its own set of terms for describing the points, lines, polygons, and pixels used to represent spatial aspects of data. For example, a unidimensional set of nonintersecting, non-branching coordinates can be variously known as a line, arc, chain, link, edge, or string. One of the most successful aspects of spatial data transfer standards has been to standardize such terminology and establish definitions for each term. Using these definitions enable vendors to match respective terminology to that of the standard.

Spatial objects as defined in SDTS.

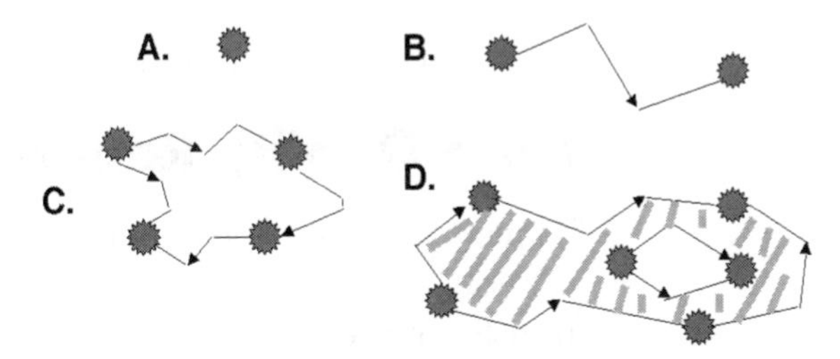

The following list contains definitions of several terms in the SDTS.

- Node (item A in the preceding illustration): A zero-dimensional object that is a topological junction of two or more links or chains, or an end point of a link or chain.

- Chain (item B): A directed non-branching sequence of nonintersecting line segments and/or arcs bounded by nodes (not necessarily distinct) at each end.
- Complete chain (not shown): A chain that explicitly references left and right polygons and start and end nodes. It is a component of a two-dimensional manifold.
- Ring (item C): A sequence of nonintersecting chains, strings, and/or arcs, with closure. A ring represents a closed boundary, but not the area inside the closed boundary.
- GT-Ring (not shown): A ring created from complete and/or area chains.
- GT-Polygon (item D): An area that is an atomic two-dimensional component of one and only one two-dimensional manifold.

As part of the implementation process, a GIS software vendor or spatial data producer would map, or find relationships, between native terms and standard terms. As a data consumer, you may or may not be exposed to the standardized terms after the data has been imported into a target system. To evaluate potential data sources, you must be familiar with terminology. As mentioned previously, this type of information can be found in overview articles and is considered part of a standard's conceptual level.

Data Quality Report

While some data consumers may find data quality reports of dubious value, if a report is included with your particular transfer project, take the time to read it. The geospatial equivalent of a "readme first" file, the report is prepared by the data producer to assist you with data assessment (i.e., determining if the data quality is adequate for your application). Depending on how rigorous your assessment techniques are and how detailed the report is, you may want to perform independent testing.

Different standards require varying amounts of data quality information and mandate that it be encoded in different ways (e.g., restricted field values; short, terse comments limited to 80 characters; or flowing text with simple formatting). When you are investigating a standard, determine its requirements for a data quality report and the form the report will take. Samples of data quality information follow.

- Sample lineage statement: "This digital line graph was digitized from a standard U.S. Geological Survey quadrangle (name, date, and scale of quadrangle indicated in SDTS identification module). The digital data were produced by either scanning or manually digitizing a stable based copy of the graphic materials. The scanning process captured the digital data at a scanning resolution of at least 0.001"; the resulting raster data were vectorized and then attributed on an interactive editing station. Manual digitizing used a digitizing table to capture the digital data at a resolution of at least 0.001"; attribution was performed either as the data were digitized, or on an interactive editing station after the digitizing was complete…"
- Sample positional accuracy statement: "Accuracy of these digital data (if not digitally revised) is based upon the use of source graphics compiled to meet National Map Accuracy Standards. NMAS horizontal accuracy requires that at least 90 percent of points tested are within 0.02" standard error in the two component directories relative to the source graphic. NMAS vertical accuracy requires that at least 90 percent of well-defined points tested be within half a contour interval of the correct value. Comparison to the graphic source is used as control to assess digital positional accuracy. Cartographic offsets may be present on the graphic source because of scale and legibility constraints…"
- Sample completeness statement: "Data completeness for unrevised digital files reflects the content of the source graphic. Features may have been eliminated or generalized on the source graphic because of scale and legibility

constraints. If the digital data underwent limited update revision, then the content will include only (1) those features that are photoidentifiable on monoscopic source, supplemented with limited ancillary source, and (2) those features that cannot be reliably photoidentified but that are not considered particularly prone to change…"

- Sample logical consistency statement: "Certain node/geometry and topology (GT) polygon/chain relationships are collected or generated to satisfy topological requirements. (The GT-polygon corresponds to the DLG area.) Some of these requirements include the following: chains must begin and end at nodes, chains must connect to each other at nodes, chains do not extend through nodes, left and right GT-polygons are defined for each chain element and are consistent throughout the transfer, and the chains representing the limits of the file (i.e., neat line) are free of gaps…"

Although the data quality information is embedded in the transfer files, the information is only useful for you, the data consumer. To read the contents of a data quality report, use a viewing utility, rather than digging through the source data files. A robust implementation will allow you to view this information prior to full import. In other cases, you may need to import the data first before you can access the information.

Attributes

Encoding attributes constitutes the most difficult aspect of spatial data exchange. Because there are so many ways to encode attributes (or descriptions of real world features), you could devote considerable time converting attributes from different source data sets into the model your application requires.

Perhaps the greatest potential benefit of a data transfer standard is to make the attribute model and set of feature types and attributes uniform. Such activity has proven extremely difficult because there is no single classification system that works for all applications. The way in which a hydrologist views the world is vastly different from that of a cartographer or even geologist. The SDTS specified dictionary is an attempt at a non-discipline specific data dictionary. It proposes a set of general feature types, and leaves refinements to individual attribute descriptions. However, because this approach has not been widely accepted by data producers, the SDTS also permits data dictionaries defined by data producers.

Different world views.

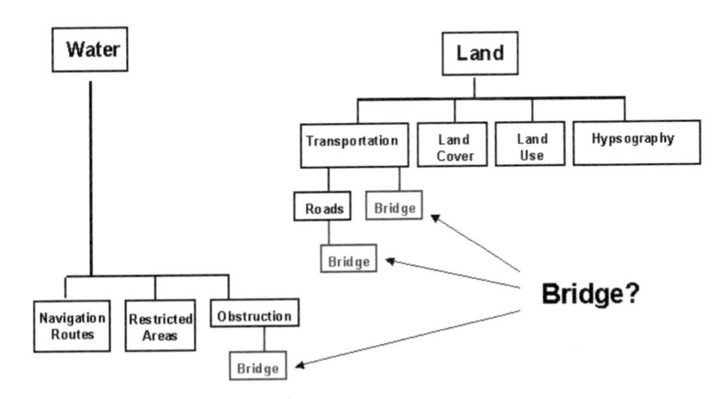

The term "bridge" is viewed differently in different contexts, as the illustration above shows. A more structured SDTS feature type follows.

- Bridge: A structure erected over a depression or obstacle to carry traffic or a facility such as a pipeline.
- Included terms: Bascule bridge, covered bridge, drawbridge, footbridge, lift bridge, pontoon bridge, overpass, pedestrian/bicycle overpass.
- Possible attributes: Bearing_Capacity, Construction_Type, Covered/Uncovered, Feature_Spanned, Width, and so forth.

An alternative approach to the "single dictionary" is developing many standardized dictionaries. The Federal Geographic Data Committee supports the development of application and domain specific data dictionaries. These dictionaries are collections of feature types and attribute sets and attribute domains pertinent to industries or professions. For example, separate dictionaries for wetlands, geodetic control, census units, cadastral units, base cartographic features, and hydrology are all conceivable, and you would expect each to contain some features that overlap. At a minimum, all data sources pertaining to a common theme could potentially use the same data dictionary, which would make combining data from different sources an easier task.

The use of spatial data transfer standards in conjunction with standardized data dictionaries will eventually reduce the range of names for common attributes. But the GIS industry is not there yet. As a data consumer, you can expect to spend time rearranging, massaging, converting, and translating attributes to prepare them for use.

NOTE: *You should first select possible data sources, and then investigate whether the sources employ a data dictionary or standard set of feature types and attributes.*

GIS Import Process

Once you become familiar with a particular data standard and have a data set in hand, you must then learn how to import the data into your system. This section addresses general steps for importing to help navigate specific procedures or vendor documentation. It concludes with troubleshooting tips and resources for getting help.

Some types of GIS software will offer direct import of a transfer standard data set, and others will not. In general, widely distributed GIS softwares support more import formats, while niche market GIS and tools support fewer formats. However, if the GIS software you plan to use does not support the specific transfer standard used by your data source, format converters are available.

Automatic Conversion/Direct Import

Direct importing is the quickest way to pull outside data into your system, unless you pursue direction data collection. All outside data you use must be imported into the target system before you can use them.

✓ **TIP:** *There are several ways to locate your system's import function. On a menu-driven system such as ArcView GIS, it may be listed as a function on the File pull-down menu, or it may be part of a File Open dialog box. If the format you plan to import is not supported, use the Help index (searching for the terms "conversion" or "import") to learn about command line converters. The Help index should also provide information on supported import formats.*

During the import process all data, including spatial objects, are decoded and placed in the appropriate files of the target system. For example, an SDTS line module record representing a complete chain imported into ARC/INFO would be converted into an "arc" object, ARC/INFO's term for a line with full topology. Georeferencing for spatial objects may be automatically established using parameters from the transfer data set, allowing you to register a particular data set to other data layers or to measure ground distances. Typically, attributes are decoded and loaded into internal structures as-is (without interpretation). However, links between spatial objects and attributes must be maintained. Data dictionary terms, definitions, and value domains are placed in tables or text files, while textual metadata are placed in text fields or a log file.

The details of the import process depend greatly on the system running the import process. While it is true that the primary objective of a data transfer standard is to prevent the loss of information, not all GIS software packages offer consistent functionality in this regard. For example, the fact that a transfer data set contains a data dictionary or textual metadata does not guarantee that the importing system can handle the information. In fact, it may only import the data it can manage and ignore other information. At present, many systems do not have data dictionary capabilities and some cannot process lengthy textual information.

The GIS import command will report the results of a conversion. This can take the form of a pop-up dialog box, log file, or series of screen messages. Always read the messages to ensure that the process worked as anticipated, and look for problems that might appear as warning or error messages.

The import process is largely an all or nothing proposition: it either works and does what you need, or it does not recognize the format. If it works but does not yield exactly what you need, investigate the advanced import options discussed in the next section. If your GIS software does not import a certain format, consider using format converter utilities. Other problems are discussed in the troubleshooting section.

Advanced Import Options

Depending on the complexity of the transfer standard and the power of the import routine, you may be able to control or customize the import process. Most of the work is done automatically, and usually the default process works exactly the way you need. If this is not the case, however, consider using advanced import options.

✓ **TIP:** *To identify advanced import options in your GIS system, click on the Options or Customize button in the Import dialog box, or look for parameter descriptions in a command line based system. Always check the documentation or on-line help to locate vital information on a specific standard.*

A transfer data set may contain multiple tiles and/or multiple data layers. Rather than loading all data automatically, you may want to select certain pieces. If your system offers this option, use the viewing utility to determine if multiple pieces are included and choose the pieces you want before you begin the import process.

In addition, spatial data organization in the transfer data set may require alteration before you can use it. For example, a vector data set might not include topological relationships (i.e., links from a line to its start and end nodes, and the polygons on either side). A raster data set may have its scan origin in the lower left corner with scans proceeding bottom to top. Rather than performing these types of reorganization as separate steps after the import process, consider including these operations as part of the import process.

✓ **TIP:** *Your GIS system may allow you to customize the import process. If direct options are unavailable, the GIS may feature a scripting or macro language you can use to write routines that automatically perform repetitive sequences.*

Converting feature and attribute data to the type you need can require significant effort. With the advent of standardized data dictionaries containing feature and attribute terms, the process is getting easier. The crosswalk feature is an equivalent term from one dictionary to another, and terms such as stream, river, and watercourse, footpath/trail, or reservoir/lake also are equivalents. As these terms become more widespread in the industry, applying a crosswalk could become part of the import process. Investigate whether the standard you plan to use contains a data dictionary that can automate attribute manipulation. If not, a macro language may serve a similar purpose.

If the native import command does not offer the control you need, you may need to consider a format converter utility.

Format Converter Utilities

Format converter utilities specialize in translating data sets from one format to another. They are essential for software that does not support the direct import of a spatial data transfer standard, and they can be helpful if you are interested in better control, more sophisticated options, or finer viewing capabilities. The large number of file formats used in the industry has created a substantial market for converter utilities, whose purpose is to translate a data set from the source system's format into an intermediate format that then can be processed by the target system. For example, to import an

SDTS vector data set into a CAD package, you could first convert it to a *.dxf* file, and then import the *.dxf* file.

To determine the candidate formats to use, compare the formats the target system can import with those the utility can export to find a match. Look for a format native to your target system with the potential for transferring the most information. For example, assume a target GIS can import data files in its native format or a simple file containing streams of XY coordinates. If the converter utility can accommodate either format, opt for the native (data file) format.

Viewing Utilities

A viewing utility is software that allows you to display, browse, and inspect the content and/or structure of a data set. Viewing utilities can be used before you import to learn about the data and after the import to measure its success. They can also be essential when an import process successfully transfers only part of the data set into the target system. In this case, a viewing utility allows you to see the data content remaining in the transfer data set, which is often the textual information such as term definitions and data quality reports.

Remember that tools for viewing a spatial data transfer are specific to the standard. A viewing utility might be part of the GIS or available as separate software. In addition, a utility can be a simple text dump or a sophisticated interactive program with spatial display.

A sample viewer text report appears below.

```
SDTS Report
Input DDF file: hy01iden.ddf
Field              Subfield  Format  Length  Data
-----                -----   -----   -----   -----
0001                                   6     1
IDEN                 MODN    A         4     IDEN
IDEN                 RCID    I         6     1
IDEN                 STID    A         30    SPATIAL DATA TRANSFER STANDARD
```

```
IDEN                STVS   A    12    1994 JUNE 10
IDEN                DOCU   A    14    FIPS PUB 173-1
IDEN                PRID   A    31    SDTS TOPOLOGICAL VECTOR PROFILE
IDEN                PRVS   A    25    VERSION 1.0 JUNE 10, 1994
IDEN                PDOC   A    17    FIPS 173-1 PART 4
IDEN                TITL   A    37    WILMINGTON SOUTH, DE-NJ/HYDROGRAPHY
IDEN                DAST   A    5     DLG-3
IDEN                MPDT   A    4     1987
IDEN                DCDT   A    8     19951031
IDEN                SCAL   I    8     24000
IDEN                COMT   A    164   This transfer requires an external
   data dictionary...
CONF                FFYN   A    1     Y
CONF                VGYN   A    1     Y
CONF                GTYN   A    1     Y
CONF                RCYN   A    1     N
CONF                EXSP   I    11
CONF                FTLV   I    14
$$$$$ END OF FILE $$$$$$$$$$$$$$$$$$$$$$$$$$$$$$$$
```

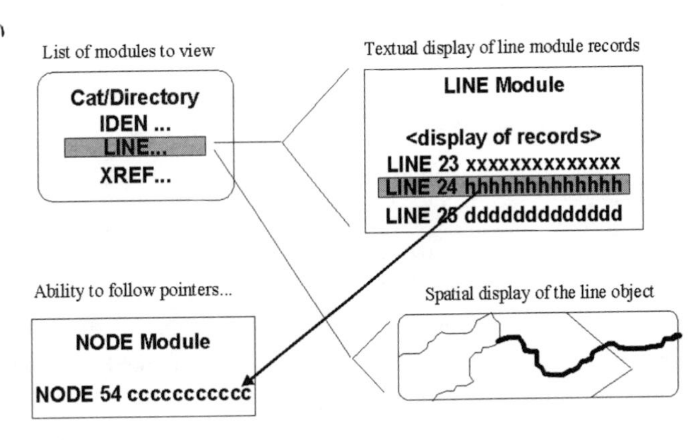

A viewer with spatial display.

You can inspect the data to identify spatial contents. You can determine the number of layers and tiles and how they relate. Before importing, you can read the metadata to assess the data's suitability. Important aspects might include positional accuracy, the accuracy of the attribution, or even the extraction rules used to collect the data set. Inspecting the data beforehand enables you to estimate the amount of manipulation the data will require (e.g., vertically integrating separate layers, reorganizing attributes, or determining how much processing is required to combine the data with another source). A viewing utility can provide information regarding how to customize the import process.

After importing the data, you have all the tools of the native system available to you to inspect the data and evaluate whether the import performed as expected. For example, if there were three tiles and two layers of data, then a full conversion would have to make all this information available. The attributes from the transfer data set would be loaded and associated with respective spatial objects. Once the data are loaded, you can review what you have, and then plan the next step.

An SDTS Import Example Using ARC/INFO

To demonstrate the GIS import command in operation, this section contains a sample import of an SDTS data set of vector data into ARC/INFO (manufactured by Environmental Systems Research Institute, Inc.). ARC/INFO uses a command line interface for its SDTS import and features a macro language to facilitate data manipulation. ESRI provides information about SDTS to help ARC/INFO users understand how to use the software properly. The ESRI white paper "SDTS: Supporting the Spatial Data Transfer Standard in ARC/INFO" (July 1995) contains supplemental information on the import process and manipulating attribute links. The paper can be accessed via the ESRI Web page, at *http://www.esri.com.* Information regarding SDTS and USGS data sets can be accessed at *http://mcmcweb.er.usgs.gov/sdts.*

SDTSINFO

To prepare for importing an SDTS transfer data set, you must prepare your workspace. Each SDTS transfer data set should be placed in a separate directory at the same directory level. If a data dictionary transfer is required, it should also be placed in its own directory at the same level called *masterdd.* Your workspace might look like this: */palth/*

project2/hydro, */palth/project2/plss*, */palth/project2/roads*, and */palth/project2/masterdd*. Within each directory, you must uncompress the files, which usually involves an unzip and/or tar utility, and is explained in the readme files provided by the data producer. Each SDTS transfer contains many files, all with the *.ddf* extension. Finally, within ARC/INFO, you must initialize the coverage to double precision by issuing the command precision double double.

To complete the transfer, you must first determine the contents of the transfer data set. The ARC/INFO SDTSINFO command provides this information. In a sense, SDTSINFO is a unique viewing utility that complements the command for performing the import, SDTSIMPORT. The information resulting from SDTSINFO is what you use to provide input parameters to the SDTSIMPORT command.

Running SDTSINFO produces a text report that scrolls on your computer screen. A listing of the transportation category for a USGS 1:24,000-scale DLG-3 SDTS/TVP data set for Alda, Nebraska, follows (line numbers are for reference purposes).

```
1)                           Arc: sdtsinfo tr01
2) Description of SDTS transfer tr01
3)
4) IDENTIFICATION::
5) Standard Identification . : SPATIAL DATA TRANSFER STANDARD
6) Standard Version. . . . . : 1994 JUNE 10
7) Standard Documentation Ref: FIPS PUB 173-1
8) Profile Identification. . : SDTS TOPOLOGICAL VECTOR PROFILE
9) Profile Version . . . . . : VERSION 1.0 JUNE 10, 1994
10) Profile Documentation Ref : FIPS 173-1 PART 4
11) Title . . . . . . . . . . : ALDA, NE / TRANSPORTATION
12) Data Structure. . . . . . : DLG-3
13) Map Date. . . . . . . . . : 1974
14) Date Set Creation Date. . : 19961114
15) Scale . . . . . . . . . . : 24000
16) Comment . . . . . . . . . : This transfer requires an external data dictionary
    from the U.S. Geological Survey, National Mapping Division, with a 4-character
    code of DLG3, version number 3.00
```

```
17)
18) CONFORMANCE::
19) Composites. . . . . . . . : Y
20) Vector Geometry . . . . . : Y
21) Vector Topology . . . . . : Y
22) Raster. . . . . . . . . . : N
23) External Spatial Ref. . . : Yes
24) Features Level. . . . . . : All non-SDTS
25)
26) INTERNAL SPATIAL REFERENCE::
27) Spatial Address Type. . . . : 2-TUPLE
28) Spa. Add. X Component Label : EASTING
29) Spa. Add. Y Component Label : NORTHING
30) Horizontal Component Format : BI32
31) Scale Factor X. . . . . . . : 0.01
32) Scale Factor Y. . . . . . . : 0.01
33) X Origin. . . . . . . . . . : 0.0
34) Y Origin. . . . . . . . . . : 0.0
35) X Comp. of Horiz. Resolution: 0.61
36) Y Comp. of Horiz. Resolution: 0.61
37)
38) EXTERNAL SPATIAL REFERENCE::
39) Reference System Name . . . : UTM
40) Horizontal Datum. . . . . . : North American 1927
41) Zone Number . . . . . . . . : 14
42)
43) GLOBAL AND DATA QUALITY MODULES::
44)
45) Module   No. Recs      Description                      External
46) ------   -------- ---  ---------------------------      ------
47) IDEN        1          Identification
48) CATD        25         Catalog/Directory
49) CATX        2          Catalog/Cross-Reference
50) CATS        25         Catalog/Spatial Domain
51) IREF        1          Internal Spatial Reference
52) XREF        1          External Spatial Reference
53) MDEF                   Data Dictionary/Definition         XX
54) MDOM                   Data Dictionary/Domain             XX
55) DDSH        45         Data Dictionary/Schema
56) STAT        23         Transfer Statistics
```

```
57) DQHL          1          Lineage
58) DQPA          1          Positional Accuracy
59) DQAA          1          Attribute Accuracy
60) DQLC          1          Logical Consistency
61) DQCG          1          Completeness
62)
63) Number of data layers: 1
64) (blank)
65)
66) Layer: (blank)
67) Theme: ROADS AND TRAILS
68)
69) Module No. Objs Object Type
70) ------ -------- ------------------------------
71) NP01     4        Point
72) NA01     163      Area point
73) NO01     600      Node, planar graph
74) LE01     762      Complete chain
75) PC01     164      GT-polygon composed of chains
76) FF01     1        Composite
77)
78) ATTRIBUTE PRIMARY MODULES:: 4
79)
80) Module No. Recs Theme
81) ------ -------- --------------------
82) ACOI     317      ROADS AND TRAILS
83) AHDR     1        ROADS AND TRAILS
84) ARDF     731      ROADS AND TRAILS
85) ARDM     154      ROADS AND TRAILS
```

The SDTSINFO command on line 1 has a single parameter. The four-character transfer prefix is found on every SDTS file that identifies the transfer file set. This command selects information from many SDTS files of the same SDTS transfer and reports on the transfer as a whole.

On lines 16, 53, and 54, you see references to an "external data dictionary." These comments indicate that you must obtain a transfer data dictionary, in addition to the transfer data set, for complete information.

On line 63, you see the number of sections of data included in the transfer. While the term "layer" is used, it can also mean adjacent tiles or vertically overlapping layers. This highlights the inconsistencies regarding terminology mentioned previously. Because this report uses the SDTS standard, it relies largely on SDTS terms, which may have different meanings from similar ARC/INFO terms. Therefore, to fully understand these reports, you must learn how various terms compare.

Line 63 tells you there is only one data layer, and lines 66 to 76 further describe the layer. If the import involves more than one data layer, the report would generate information to help you decide which layers to import.

The SDTSINFO command is convenient because it examines many SDTS modules and summarizes the contents of the SDTS transfer. If you need to examine an individual SDTS module (or file), use the SDTSLIST command to display a module's contents as a text report. If you issue the SDTSLIST command on the *IDEN.DDF* file you would see the same information found in lines 4 through 24 of the SDTSINFO listing. The command with its parameter would be SDTSLIST TR01IDEN.DDF. The data quality report (found in files *DQPA.DDF*, *DQAA.DDF*, *DQHL.DDF*, and *DQLC.DDF*) and data dictionary information (*MDEF.DDF* or *DDDF.DDF* and *MDOM.DDF* or *DDOM.DDF*) can be very pertinent. The SDTSLIST and SDTSINFO commands work off of the SDTS transfer files and can be run prior to or following the import.

SDTSIMPORT

The SDTSIMPORT command loads a single layer at a time into ARC/INFO. Because there is only one data layer in this example, the import command will only need to be run once. Syntax and usage for the SDTSIMPORT command follow.

```
SDTSIMPORT <in-transfer-prefix> <out-cover> {out-point-cover}
{layer-name} {DD or DROP_DD}
```

<in-transfer-prefix>: The four-character prefix on an SDTS **.ddf* file.

<out-cover>: The new arc cover name.

{out-point-cover}: The new point cover name.

{layer-name}: The SDTS data layer name you will import.

{DD or DROP_DD}: Loads or disregards data dictionary information.

The *layer-name* parameter is where you specify the names from the SDTSINFO report. When this parameter is not specified, the default is to load the first data layer in the transfer. Therefore, it is possible to miss layers if you choose the default; you might not realize layers are missing. The DD parameter indicates whether to load the data dictionary. If the SDTSINFO report tells you the dictionary is external, you must retrieve it separately. If the dictionary is internal, it is included with the data set. You only need to load the dictionary once during a transfer; it can be loaded with any data layer.

The following list shows the scrolled messages that appear on the computer screen when the SDTSIMPORT command is invoked.

```
1) Arc: sdtsimport tr01 transp
2) Importing ARC coverage transp from SDTS/TVP transfer tr01.
3) Importing layer <blank>
4) IDENTIFICATION::
```

[See SDTSINFO report, lines 4 through 17.]

```
5) CONFORMANCE::
```

[See SDTSINFO report, lines 18 through 23.]

```
6) Features Level. . . . . . : All non-SDTS
7) Total Features: 1526
8) Nodes: 600
9) Arcs: 762
10) Polygons: 164
```

Line 1 is the command showing which parameters were entered. Note that the second parameter is the cover name *transp*. Lines 2 and 3 inform you of the data layer being loaded and the kind of transfer (*SDTS/TVP*). Warning or error messages would appear after line 6. Lines 7 to 10 report the total number and types of objects created in ARC/INFO terms as a result of the import. Compare these numbers and terms to those listed in lines 73 to 75 of the SDTSINFO report. For example, line 74 indicates the SDTS term "complete chain" translates into the ARC/INFO term "arc" on line 9 of the SDTSIMPORT report. To determine whether georeferencing information was automatically imported, issue the DESCRIBE command.

```
Arc: describe transp
Description of DOUBLE precision coverage transp
FEATURE CLASSES
```

[Unrelated details appear on the screen.]

```
COORDINATE SYSTEM DESCRIPTION
Projection UTM
Zone 14
Datum NAD27
Units METERS Spheroid CLARKE1866
```

The data have been successfully loaded into ARC/INFO. You can issue ARC commands and use software tools to work with the data. You can even write your own macro programs using the ARC macro language (AML) to help automate the process and work with attribute structures. Recovering the correct spatial object and attribute links can be tricky. If not joined correctly, you will get incorrect attributes without warning messages. Each spatial object can have one or more sets of attribute lists. In ARC/INFO, the *<cover-name>.rel* file shows the different relate tables and records cross-referenced to the appropriate tables. If there is no *<cover-name>.ajoin* INFO file, then each element has only one attribute link.

NOTE: *This chapter addresses only the issues data consumers have regarding transfer data sets. Data export using a spatial data transfer standard, which requires far more explanation, is beyond the scope of this text.*

Troubleshooting

Murphy's law appears to thrive wherever computers are used, and GIS data conversion is no exception. The following guidelines discuss common problems that plague spatial data transfers and suggest methods for solving them. Sources for additional help are also provided.

Diagnosing the Problem

How do you know when the import process performed incorrectly? Typically, the system will provide a warning or error message indicating the problem. If your software hangs (stalls, or appears inactive) or crashes during the import process, this indicates a problem as well. The data set may sometimes appear incorrect after you complete an import. For example, there may be large chunks of spatial data missing, or the attributes attached to spatial objects may not make sense (e.g., if a line is attributed as a lake). These problems may be caused by the import software, data set, or both.

Imports can fail if appropriate versions or profiles are not implemented, users do not follow software instructions, or bugs exist in the software. The data set can cause problems if its files are incorrectly formatted or encoded or if they are corrupted. Some problems are easy to diagnose, while others may require that you contact the software vendor or data producer.

Common Problems

The following tables may help you classify your problem, determine the cause, and in some cases identify a solution.

Error message: Format unknown/not recognized

Problem	Possible cause	Action
Software	Standard not supported.	1. Obtain another format from data producer. 2. Inquire about a software upgrade with support. 3. Use a format converter utility.
Software	File extension incorrect.	Check import documentation to determine if a specific file extension is required (e.g., *.ddf for SDTS or *.jpg for JPEG).
Software	Preprocessing needed to unpack one or more files.	If the file has a .tar, .gz, .z, or .packed extension, use a tar/zip utility first.
Data set	File corrupted.	See the "Error encountered while importing" table that follows.

Error message: Profile/version not supported

Indicates that a specific profile or version of a standard is not yet supported.

Problem	Possible cause	Action
Software	Data model incompatible (e.g., vector data in a raster system).	1. Check with data producer to obtain different data model type. 2. Consider using different software.
Software	New profile or standard version.	Ask vendor about software upgrade or plans to implement upgrade.

Problem: System hangs or crashes

Problem	Possible cause	Action
Software	Bug in import process.	1. If problem occurs on more than one data set, then contact software vendor. 2. If problem occurs on a single data set, contact data producer.

Data set	Incorrectly encoded; import starts but never finishes.	If problem occurs on more than one system, contact data producer.
Other	System problem.	Restart computer and try the import again; if problem repeats on the data set, try a second data set. If problem occurs with both data sets, contact data producer.

Error message: Error encountered while importing

Problem	Possible cause	Action
Data set	File corrupt, bad media.	Try reading files from removable media (e.g., floppy, CD-ROM, tape) to hard drive. If unsuccessful, obtain a new copy.
Data set	File corrupt, download error.	If an FTP or Internet download, determine whether appropriate settings were used (i.e., binary versus ASCII).
Data set	File corrupt, unpack error.	If untar or unzip used, determine whether appropriate options were selected (e.g, disabling automatic LF/CR during unzip).

System error or warning: Can't locate information...

Problem	Possible cause	Action
Data set	Data set missing information.	Contact data provider to determine if you are missing part of the delivery package (e.g., data dictionary).
Data set	Data set encoded incorrectly.	Contact data provider for updated/ corrected information.

Error message: Invalid or missing parameters

Problem	Possible cause	Action
Software	Parameters in wrong order.	Check system documentation or on-line help for information about parameters.
Software	Invalid parameter values.	Check system documentation or on-line help for information about parameters.

Data inspection: Data missing or obviously incorrect

Problem	Possible cause	Action
Software	Parameters incorrect.	1. Check whether all tiles and layers were loaded as intended. 2. Check whether parameters work as expected.
Data set	Data set encoded incorrectly.	Contact data provider for updated/corrected information.
Data set	Attribute linkages incorrect.	Double check that you are using correct database procedures to associate spatial objects and attributes.

Where to Find Help

If you are unable to resolve the problem, there are a number of ways to seek assistance. The first place to look is documentation or on-line help that ships with the software. If you subscribe to a technical support service, this would be the next best option to get help if you pay a flat rate. If you pay per call or have no technical support plan, try the free resources listed in the next section first. Before seeking help, have the following information available to expedite your inquiry.

- Name of software and hardware you are using, including version information.
- Name of your data set and data producer.
- Steps you followed.
- Error or errors you encountered and the points at which they occurred; include all system messages.
- Research and troubleshooting steps you have taken.

Suggested Places for Help

GIS Related News Groups

If you are already active in Internet news groups, consider seeking assistance from your peers. Because vendors and data producers are occasionally active in news groups, this can be an effective way to access a very broad knowledge base. In addition, many news groups have frequently asked question (FAQ) documents and archives of past messages. Your question may have already been asked and answered. For import questions, try the news groups *comp.infosystems.gis* and *sci.data.formats.*

E-mail Lists

An e-mail list is comprised of e-mail users who subscribe to a given mailing list. Mail sent to the list is forwarded to the private mailboxes of everyone on the list, and people can choose to respond to the list or an individual. The mailing list dedicated to transfer and translation issues is *gistrans-l.* Other GIS related mailing lists are *maps-l, nsdi-l, geonet-l, nsgic-l, gis-l,* and *gisbus-l.*

E-mail to Vendors or Data Producers

If the vendor or data producer offers direct e-mail technical support, send your message directly to the address provided.

Web Pages for Vendors or Data Producers

Many companies put extensive information on line to make it more accessible. Check Internet Web pages for product and data information. Use one of the many search engines to locate additional sources of information.

Acknowledgments

I wish to gratefully acknowledge the contributions of Gladys Conaway and George Miller in providing the material for the import example. Their expertise and willingness to help are greatly appreciated. I would like to thank the U.S. Geological Survey for supporting my contributions to this effort.

References

Moellering, H. and Hogan, R., editors. 1997. *Spatial Database Transfer Standards 2: Characteristics for Assessing Standards and Full Descriptions of the National and International Standards in the*

World. International Cartographic Association. Pergamon, Elsevier Science.

National Institute of Standards and Technology. 1994. *Federal Information Processing Standard Publication 173-1 (Spatial Data Transfer Standard).* U.S. Department of Commerce.

Spatial Data Transfer Standard Information Site, on the World Wide Web, accessed using the URL *http://mcmcweb.er.usgs.gov/sdts.*

Section 6: Quality Control/ Quality Assurance

This final section suggests methods for proper quality control and quality assurance procedures in order to maximize data quality.

Quality Control/Quality Assurance

Edward Kura

One barrier that prevents new GIS projects from becoming successful is user mistrust of GIS data. For many reasons, some users cannot accept data in a GIS as representative of paper predecessors. GIS data do not look the same as they appear on traditional paper sources, even when plotted in a similar format. Users generally expect and compensate for inaccuracies and deviations on paper maps, often without realizing it. However, when viewing the same data in a GIS, quirks and inaccuracies in data are equated with the system housing them.

To overcome this problem and persuade users to accept and use a GIS, you must convince them that the GIS data are at least as accurate as data in paper systems. In many cases, users expect GIS to be more accurate than paper sources. They view the creation of a GIS as an opportunity to correct errors they have compensated for over the years. Because computers do not make mistakes, users believe GIS data should be perfect. However, while most GIS professionals will agree that computer errors are infrequent, data in a GIS are rarely 100% accurate.

Consequently, one of the most important goals in a GIS implementation must be to gain user acceptance. A key to that acceptance is the accuracy of information that the GIS makes available. Users must believe that a GIS can provide accurate, useful information in an accessible format, one of the hallmarks of modern GIS. The time to ensure that your data are as accurate as possible is when you convert them. To do so, you must implement quality control and quality assurance procedures.

Attribute versus Graphic Quality

GIS data quality can be divided into two equally important parts: attribute and spatial accuracy, each of which is described in detail in the following sections.

Attribute Accuracy

In a GIS, attributes are defined as feature characteristics. In the database design, you define a set of legal attributes for each feature type, and every feature of the same type will have the same set of attributes. For example, in a water distribution system GIS, a water pipe would be a feature and its attributes might be material, year installed, depth below ground, and pipe size. All water pipes within the GIS would have the same set of attributes. However, different pipes would have different values for the same attribute. For example, some pipes would have a pipe size of 8", while others would have a pipe size of 16".

Attribute quality is measured in relation to the source documents used for the data conversion. Each attribute value in the GIS database should correspond to an entry or notation on the source document used for the conversion. Where information exists on original sources, it should be accurately reflected in the database. Where data are missing, the attribute value should either reflect a predefined placeholder value (default) or be left blank. You should decide how to handle missing source data during the database design phase. If you choose default values, you must define default values for every attribute of every feature type at the outset. Default values are preferred over blank fields for quality control purposes.

If an attribute that was supposed to be filled in is missed during conversion, it is counted as an error and must be corrected. Blanks used as default values make quality control and quality assurance problematic because it is difficult to determine whether the blank indicates an actual default value or a missing attribute value. Filling every attribute field with real values or predefined defaults removes this question from quality control and assurance processes.

Some GIS databases may require that attribute fields be left blank when no source data are available. In such situations, default values can be defined that are clearly outside the possible set of real attributes. For instance, a water pipe material called "default" could be defined as part of the database design. After the data conversion is complete, all water pipe material attributes containing the characters "default" would be replaced with a blank field. In this way, the quality control and assurance processes are simplified and GIS database design requirements are satisfied.

What About the Data Source?

GIS data are only as accurate and reliable as the sources from which they originate. In cases where attribute values are available from more than one source, the sources may disagree. If this occurs, it may be helpful to end users to know the source of an attribute value. This can be accomplished through the use of additional attributes known as source codes. For each attribute for which you want to identify the source, a corresponding source code attribute is added to the feature in the database design.

Each possible data source for every attribute must be assigned a code. A single number or letter is usually adequate. These codes are then used to populate the source code attributes. The use of a single character for source codes will help minimize the size of the final database. If you can justify the additional storage space required, non-coded values such as "Inventory Map" can be more meaningful for end users.

Default attribute values may be confusing to end users if source code attributes are not included in the database, particularly if common values are assigned as defaults. When the user receives the results of database queries, he will not know if the "common" values are actual or default values, because actual values were not known when the database was populated. The following table shows how codes might be assigned to sources.

Assigning source codes to a GIS database

Source code	Data source
1	Primary data source
2	Second data source
3	Third data source
4	Fourth data source
5	Default value entered

Spatial Accuracy

There are several aspects to spatial accuracy and graphic quality, and each must be addressed in a complete quality control/quality assurance strategy. Elements of spatial accuracy are listed below.

- **Absolute spatial accuracy.** The placement of features relative to their actual location on the Earth's surface.
- **Relative spatial accuracy.** The placement of features relative to base map data and other features in the GIS.
- **Legibility.** Clarity of the data presentation.
- **Proper symbology.** All features must be shown with correct symbols, at the indicated size, rotation, and so forth.
- **Other project specific criteria.** Each project has a unique set of requirements that must be met and quality control checked.

You should define each element listed above as part of the quality standards for your conversion project. As data accuracy and quality requirements are tightened, standards for each element must be defined in greater detail. As accuracy requirements approach 100%, costs can rise dramatically. It has been said that 95% of the cost of data conversion can be incurred in insuring the accuracy of the data from 95% to 100%. In other words, 95% of your budget can be spent on the last 5% of quality. In brief, you must find a balance between available funds and accuracy requirements.

Defining accuracy requirements is a difficult but important part of GIS data conversion. While everyone would like to have 100% accuracy, it is often not the most cost effective approach. Remember that a GIS database should not be static. Its information and capabilities should be continuously updated and enhanced, and one possible enhancement is improving data accuracy. It is not necessary to reconvert all data to improve accuracy. As new data become available, they can be added to the database. Remember that new data will improve in accuracy as technology advances, making it more cost effective to collect increasingly accurate data. This is particularly true with spatial data because the cost of GPS data collection continues to decline as accuracy improves. As new data are added, the accuracy of the overall database will improve.

The level of accuracy of source data must be considered when you define spatial accuracy standards for the project. Including a step in the conversion process to examine and improve data quality on source documents may be desirable. The step can be included as part of the document scrubbing described in Chapter 14, "Keyboard Entry of Attribute Data."

Organizations will frequently use existing base map data as a starting point for GIS applications. The accuracy of base map data defines an upper limit for the spatial accuracy of the entire GIS. This leads to the issue of absolute versus relative accuracy. If data are placed in the GIS with absolute accuracy (at georeferenced coordinates), they may appear inaccurate relative to base map data. If the data are placed with relative accuracy (relative to the base map), they may lose accuracy relative to georeferenced coordinates. One solution to this dilemma is to place the data in the GIS with relative accuracy, but store collected georeferenced coordinates as feature attributes. Using this approach, georeferenced coordinates may be used in the future if and when the accuracy of base map data improve.

Legibility is another aspect of quality that must be defined at the beginning of a data conversion effort. Because GIS data may be presented at any scale, your definition of legibility should include the scale at which the data must be legible. Symbol and annotation definitions should include appropriate sizes for hardcopy and computer legibility. Sizes appropriate for a scale of 1:100 may be inappropriate for a scale of 1:2,000. GIS project specifications should state the display scales to be accommodated.

Proper symbology is also part of GIS project specifications. This definition should include the symbols used for each feature, fonts and font sizes, colors, and which features should be displayed within different scale ranges. For instance, while it may be important to see all water valves on a display at 1:100, at 1:2,000 valves would distract you and possibly obscure other features.

Quality Control versus Quality Assurance

It is important to distinguish between quality control and quality assurance in GIS data conversion. The following sections define each term and discuss where each process belongs in the overall GIS data conversion effort.

Quality Control Definition and Philosophy

Quality control is the process of monitoring the quality of the data being loaded into the GIS database, and taking corrective action to ensure the data meet the project's defined standards. More importantly, quality control needs to be a state of mind because data quality should be built into the conversion effort as a part of all processes, not just a step where accuracy and quality are checked.

Each part of the conversion process must be designed with quality in mind. Rather than depending on a quality control step to find errors after they occur, design the process to anticipate errors and plan methods to avoid them. While quality control checking and correction are important, it is more critical to prevent errors from occurring in the first place. In addition to the costs you will incur to find errors, there are costs to correct them and verify that corrections are accomplished properly. Basic tips for avoiding errors before they are introduced follow.

Transcription Errors

Avoid transcription errors by having the person who finds data also enter them into the computer. If the person who finds data also records them for another individual to enter them into the computer, errors can be made by the second person in the process. Under a complete quality control program, adding a separate data entry step means that you must also add a step to check the information before it is keyed in. Each step increases project cost.

Structuring Assignments

Structure work assignments so that staff members can become experts in one part of the process. Data conversion requires several skills (e.g., data interpretation, keyboard entry, data research, digitizing, and checking). Identify each staff member's strongest skills and assign work appropriately.

Rotating Assignments

Rotate work assignments periodically. While people may work more efficiently on some assignments than others, if monotony sets in, errors will follow. Rotating assignments will also expose staff to the entire conversion process. The more they understand, the better they will be at executing the process and making constructive suggestions to improve it.

Allow Feedback

Listen to staff suggestions. Even the most carefully designed conversion process can be improved by the individuals performing the work. However, evaluate and test proposed changes, so that they do not adversely affect other steps in the process or hurt quality.

Separate the Work from Quality Control

Never allow staff members to check the quality of their own work. It is quite likely that they will repeat mistakes they made when doing the work originally. While this may prove inconvenient at times, it is extremely important.

Peer Review

Always try to have a colleague or peer review your process and changes you propose. Another pair of eyes may see problems that you do not.

In a data conversion project, a separate place exists for quality control steps, and they are very important. These should take place after major conversion tasks such as digitizing are completed. Think of quality control steps as the second line of defense against errors, after overall quality control has been built into the conversion process. These tasks should be assigned to staff members who understand the conversion process and the data being converted.

Another important premise is that errors should be detected and corrected as early in the process as possible. The longer an error exists, the more difficult and costly it will be to detect and correct. Staff should be encouraged to find errors. At the same time, create an atmosphere that conveys the necessity of correcting errors, does but not blame individuals. If you determine that one person repeatedly makes the same errors, provide the individual with more training in the problem area. It is very important that your conversion staff not fear errors. They should be rewarded for finding and correcting them as early as possible. Management must recognize that people will always make mistakes.

Quality control consists of two phases. The first phase uses software to perform checks on converted data. It checks to ensure that attribute values are valid using a predefined table of allowable attribute values. Depending on the type of attribute, it may check for specific values or that a value falls within an acceptable range. Data that do not meet specified criteria are flagged as errors. The software may also check for relationships between attributes. For instance, if the material is cement, the size must be between 8" and 24". While quality control software will check data and flag errors, conversion staff must evaluate the results and correct errors found by the software. The software should be rerun after corrections until no errors are found or an applicable standard is met. If more than one quality control program is used, each should be rerun after a round of corrections to ensure that no errors were accidentally introduced during the correction process.

The second phase of quality control is manual checking. During manual checking, plots of converted data are compared with source documents to verify that data have been converted properly. Side-by-side checks of plots and sources should be performed to verify data completeness. Where practical, plots should be produced at the same scale as the source documents. These plots should be overlaid on the sources using a backlit table to verify that digitized data are placed correctly. Errors found on the plots should be marked. Ideally, corrections should be made by the person who did the original work. This process will allow her to learn from her mistakes and should improve data quality over the course of the project.

✓ **TIP:** *Checking plots can be customized to help make spotting errors obvious. Use color and varied symbology to reveal illogical data patterns. For example, plot water lines with the line thickness based on the diameter "size." Water typically flows from large pipes to smaller ones. In this way, an incorrect and illogical size could be spotted easily (e.g., a 1" pipe feeding a 12" pipe).*

Attribute fields should also be checked as thoroughly as possible. Ideally, all attributes should be checked. If this is not feasible because of system or budgetary constraints, check as many attributes as possible, emphasizing the most important ones. Rank attributes based on your GIS applications and input from end users.

The most important aspect of a quality control program is attitude. Quality must be an attitude present throughout the conversion process. Staff must be encouraged to report errors and problems without fearing repercussions. Quality will be affected positively if the staff assumes responsibility for the quality and accuracy of the converted data.

Quality Assurance Definition and Philosophy

Quality assurance is a final check of converted data before they are loaded into the GIS database. The two primary goals of quality assurance are to monitor the final quality level of the converted data and ensure that the overall quality control process is functioning properly.

Details of the quality assurance program should be laid out at the beginning of the data conversion effort. These should include the types of software and manual checks that will be executed and the level of accuracy required. The level of accuracy will define the number of errors allowed before a data set is rejected. The American Society for Quality Control publishes American National Standard, Sampling Procedures and Tables for Inspection by Attributes. This document presents an acceptance sampling system with associated tables for determining sample size and allowable number of errors (rejects). While the document is geared toward the manufacturing environment, where raw materials and final product must be inspected and either accepted or rejected, the statistical information included can be valuable in setting up a quality assurance program for GIS data conversion.

Quality control monitors the accuracy of the converted data throughout the process. Depending on the software being used and the final GIS platform, there may be data translation steps included in the conversion process. While these translations are generally reliable, they can introduce errors in the data after a particular step in the quality control process is complete. These errors should be detected by the quality assurance software.

The quality assurance process uses the same phases as quality control: software checking and manual checking. However, in quality assurance, only a subset of all attributes is checked manually. A random sample of attributes should be selected and checked for accuracy against source documents.

If possible, quality control software programs should be run against the data again as a part of the quality assurance process. This will ensure that no errors have been introduced to the data since the last round of quality control checks. Because these checks are performed by software, it will probably be cost effective to run them against all converted data. It may also be possible to develop software routines to check the data in their final format. These routines may take the place of quality control programs as part of the quality assurance process.

Errors detected by all quality assurance checks should be investigated. Random errors are not significant if they do not indicate a trend and do not exceed the allowable number of errors defined under the quality assurance guidelines. Ideally, all errors found during the quality assurance process should be corrected, even if the number of errors is within the allowable limit.

All data sets containing more than the allowable number of errors fail the quality assurance process. All failing data sets should be run through all quality control checking again, and all errors corrected. Then a new round of quality assurance checking should be conducted. Repeat this process until the data set passes.

In a well-designed conversion process, a failure at the quality assurance stage usually indicates a problem in the conversion process. The problem could be related to the data themselves, where standards or allowable values may have changed without the proper modifications being made to the software. The problem may be that conversion staff needs additional training. Whatever the problem, attend to it immediately.

Data Acceptance Criteria

In a perfect world, all converted data would be 100% accurate. However, while you may strive for a perfect world, you also must establish accuracy guidelines that are acceptable today.

Define an Achievable Level of Quality

Data acceptance criteria must be realistic in terms of the level of accuracy that they set. It is almost impossible to achieve 100% accuracy, and most applications of GIS data do not require it. In fact, most GIS data conversion projects start with imperfect source data. During the conversion process, errors and anomalies can be identified and corrected. This actually improves the data and more than compensates for errors introduced by the conversion process. Loading data into a GIS frequently permits viewing the data in ways that were formerly impossible, enabling users to locate problems and improve the data.

NOTE: *"Cleansing" a data set is often a welcome side effect of the data conversion process.*

Converted data usually include multiple attributes for each feature type. These attributes can be classified by importance in terms of how the data will be used. Attributes you know will be used in GIS applications should be considered required and converted for each facility. If they are missing, research should be conducted to determine correct values. Attributes not considered required can be left blank or populated with default values. These non-required attributes can be called optional attributes.

When defining quality assurance standards and guidelines, different standards can be set for required and optional attributes. More stringent accuracy standards for required attributes would ensure that those attributes needed for GIS analysis are immediately available. Attributes converted for potential future applications could be converted to a lower degree of accuracy. This is one effective tool for controlling the costs of GIS data conversion.

Conversion costs must also be considered when establishing data acceptance criteria. Given enough time and resources, all data sets can be 100% accurate. However, most conversion projects do not have the funding to perform all research necessary to achieve complete accuracy. Data acceptance standards must be established that can be achieved within the schedule and budget of your project. In many areas, data can be improved over time as the GIS is used.

Define Criteria for Measuring Quality

Now that you have defined a level of acceptable quality for your data conversion project (e.g., 95% accuracy), you must define how quality will be measured. What checks will you use to determine whether the minimum acceptable quality level has been achieved?

There may be several aspects of data quality being measured in your conversion project. One should be graphic data quality, which is focused on spatial accuracy, legibility, and symbology. Of these three, only symbology can be measured definitively: a feature is either displayed with the correct symbol or not. Spatial accuracy and legibility are subject to interpretation and more difficult to define.

The degree of spatial accuracy achieved during data conversion is determined primarily by the data collection techniques used and the accuracy of the base map data selected for the GIS. Keep the following points in mind when establishing spatial accuracy requirements.

- Do not require accuracy greater than the sum of the possible error of the base map and the data collection technique used. For example, if the base map is accurate to within 5', and the GPS system used to collect field data is accurate within 3', then the data in the GIS could be up to 8' off their true position. In this case, you should not require that any spatial data be less than 8' from their true positions.
- If data are being digitized from paper sources, be sure to take into account the accuracy of the digitizing tablet; the digitizing process itself may introduce error. Consult the technical specifications of the digitizing hardware when selecting hardware and establishing accuracy guidelines.
- Plotted output type also can contribute to data accuracy. Paper can stretch and contract substantially, depending on plotter type, temperature, humidity, and handling techniques. Rolling up paper plots tends to stretch them in the direction rolled. Stable based media such as Mylar can help alleviate these problems, but are typically expensive compared to paper.

Legibility can be difficult to define. Something that is legible to one person may be very difficult for another to understand. In mapping and GIS applications, legibility is also related to display scale. Text labels that are clear and easy to read at one scale may be indistinguishable at another scale.

Consider the following issues when establishing legibility criteria.

- Define all output scales at which text should be legible, and the text sizes to be used at each scale. Different fonts may be used to clarify text at different scales.
- Define the data layers that will appear on top and therefore be allowed to obscure other data layers.
- Specify where arrows can and/or should be used to indicate which feature a piece of text refers to.
- Specify the text and symbol sizes to be used at different output scales.
- Consider omitting certain text labels in congested areas. Establish a hierarchy to determine which text should be displayed where on different overlays. Different standards may be required for different display scales. In addition, different standards may be required for plots and on-screen displays.
- Various output scales may be required by different end users, and each user may have unique requirements for viewing and analyzing the data.

A significant proportion of quality control and quality assurance can be performed on attribute data in an automated fashion if you document the relationships that exist between different attributes. For example, when converting water distribution system data, it could be that no water pipe over 24" in diameter is constructed of PVC. Software checks can then be run on the database to detect pipes with a size greater than 24" and the construction material PVC. Pipes meeting these criteria should be flagged as errors in the database. Attribute relationships are called logical relationships. You should define as many logical relationships in your database as possible, and use them in quality control/quality assurance checking.

✓ **TIP:** *Remember that logical relationship checking will find errors in attributes. However, it does not ensure that attribute values are correct. Such checking only* *verifies that values are acceptable within defined criteria.*

While the pipe with a diameter greater than 24" may not be made of PVC, there may be two acceptable construction material values for pipes of that size. Logical checks will only ensure that the value is one of the two, not that it is correct. Logical checks should be followed by appropriate checks of attribute data against sources.

Establish Procedures for Quality

Quality does not occur spontaneously; it is the result of careful planning and meticulous execution of quality control and quality assurance plans. Thorough quality control and quality assurance procedures are necessary to ensure a consistent level of quality throughout the data conversion effort. These procedures must be continued (although perhaps in a modified form) as part of long-term database maintenance. Creating a high quality GIS database is a major accomplishment. Do not compromise the data with shoddy maintenance procedures.

What Types of Checks When?

As you know, quality checking is divided into two types: manual checks and automated (or software) checks. Each type is more effective at a specific point in the process. You must balance the use of manual and software checking to maximize accuracy and cost.

When to Use Manual Quality Checking

Manual quality checking is labor intensive, and therefore tends to be expensive. However, in many cases, it is the only way to check data that have been entered into the database. Manual checking involves having a person inspect the data to verify that they exactly reflect the information on source documents. The high cost of manual quality checking may dictate that you check only required attributes. If so, use automated methods to check optional attributes.

Manual quality checking is also necessary for attribute data that are not restricted to a predefined list of values, such as a "comments" field.

Manual quality control checks are the only way to check graphics and spatial data. Software may check to see if a feature is displayed using the correct symbol, but people must determine if the symbol is placed correctly in relation to the base map and other converted features.

Data considered "mission critical" should always be checked manually and with the use of software. In addition, spot manual checks of such data are worthwhile backups to ensure that automated checking programs are working as intended. Manual checking is also effective when converting small amounts of data, such as maintaining data converted previously.

When to Use Automated Checking

Automated quality control checks are very cost effective for performing cursory or "sanity" checks on converted data. When designing software to convert data, you can program logical checks into the routine to detect and correct errors as they occur. This technique is particularly effective when applied to data collection performed by field crews using portable computers. Field data collection is expensive, and sending crews out repeatedly to correct errors can make the technique cost prohibitive. Applying automated software checking when data are entered into the portable computer will help eliminate many errors. If data do not pass logical checks as values are entered, the software notifies the operator, who can immediately correct the error.

Some GIS databases are constructed using existing digital data from a variety of sources. Data from these sources are simply converted to the required format and loaded into the GIS database. Automated quality control checks can be efficient for ensuring that automated data conversion is not introducing errors into the process. Software can be written to compare converted and source data and report discrepancies.

Define Detailed Quality Control/ Quality Assurance Procedures

All quality control and quality assurance procedures should be clearly defined and documented. A key factor in every successful quality program is the use of standardized procedures. All staff members should use the same procedures when performing the same tasks. Because the procedures are developed to detect errors, they must be followed by everyone.

Direct staff members who think that they have better methods for performing certain tasks to suggest that standard procedures be changed. After a thorough evaluation, the change is implemented for use by all staff. While suggestions for improvement should be encouraged, make it clear that suggestions should be tested only in a controlled situation that can be carefully monitored, rather than within the regular production process. Testing new procedures without consulting project managers may harm other parts of the process.

All data conversion staff members must also take responsibility for the quality of their work. Having a personal connection to the project will likely make them more conscientious. While staff motivation is generally discussed in the context of reducing conversion costs, it is equally important for ensuring quality.

Documentation of Quality

Producing a high level of data accuracy and quality should be a primary goal of every data conversion project. Will all end users of the GIS database believe you when you explain that the data meet accuracy requirements? Such a scenario is unlikely. You must document of the quality control and quality assurance processes to demonstrate that data meet the required standards. Documentation may also be required in the future when new users are brought into the process or when you enter into data sharing agreements.

Quality Control/ Quality Assurance Log Forms

Each step in the quality control and quality assurance processes should be recorded in a log. You may create specific forms and store them in a three-ring binder, use another form of paper record keeping, or the log may be in digital form. If the log is kept digitally, it should be printed periodically and filed as a part of the permanent project record. At a minimum, the log should record the following information.

- Quality control/quality assurance process being logged
- Individual(s) who performed the checking
- Date of the work
- Results of the checking, including the number of errors found, and how many features and/or attributes were checked
- Name or designator of the data set being checked

In addition to documenting the process, logs can also be tools for monitoring staff performance during the data conversion effort. If quality control and assurance logs are cross-referenced with logs of the data input processes, staff who require additional training may be identified. Individuals who consistently make the same mistakes often need more training, and providing assistance in the right environment can turn a marginal performer into a top performer.

Automated Reports and Logs

The best quality control/quality assurance software is useless unless it produces output that informs the data conversion staff of results. Results should be in the form of reports, which can be stored digitally and used on screen, printed, or both. Printed reports are handy because you can write notes on them and use them to research data errors and conflicts. While saving paper is noble, do not compromise the efficiency of your operation just to save paper.

Printed reports may also serve as a "paper trail" of the data conversion process. If an end user needs to know where an attribute originated, it may be necessary to review the reports to locate notes on data research. Being able to conduct such detailed backtracking can significantly bolster the end user's confidence in the data. The importance of user confidence cannot be overstated. If end users do not have confidence in the data, they will not use them, even if the database is 100% accurate.

Reports generated by automated checking programs should include the following minimum features.

- Name of the data set checked
- Date the program was run
- Name of the technician running the program and correcting errors
- Descriptive listing of each error found
- Summary of results

The name of the technician running the program and correcting errors might be filled in by hand on a printout of the report as work is completed. It may also be useful to record whether the check is in the first, second, or third round. At the very least, a printout of the final run of the program showing that no errors were detected should be kept as a permanent project record.

Logs that record who performed each step and when may also be kept as a part of the data conversion process. The log could be expanded to compare time spent on a project to the budgeted production rate, if you need to measure the efficiency of the process on a continuing basis. A database application could easily be written to record this information and provide status reports.

Problem Resolution

One of the results of the quality control process is documentation that outlines where errors exist in the converted data. Some errors are simply mistakes in keying the information or digitizing it in the wrong place. However, in some cases, the data have been converted according to the source, but they still do not pass the quality control programs. These errors require research to determine why the data do not conform to established standards. This process is known as problem resolution.

The scope of the problem resolution process for data conversion should be defined at the beginning of the project. A problem resolution process should be defined for each source document. For instance, where field data collection is being used as a source for attributes, you must decide whether it is cost effective to send crews back to the field to verify attributes that fail the quality control programs. The crew may have made a mistake the first time, or the data may be exceptions to rules established for the project. If you determine that funds are not available to send the field crew out again, consider adding a note to the feature as an attribute. All notes could be compiled in an exceptions report generated at the end of the project, and then researched as time and funding become available.

Aside from its quality management elements, problem resolution is one of the most valuable aspects of a GIS data conversion program. Problem resolution is the process that actually improves the data, increasing the value of the GIS. If problem resolution is well thought out, your GIS database will likely be far more than an electronic version of old paper data sources.

There are two common approaches to problem resolution. In one approach, it is invoked as soon as a data error or conflict is identified, and the solution is applied as early in the conversion process as possible. This approach can result in problem resolution being applied throughout the data conversion process. While it is generally best to identify and solve problems as early as possible, this approach might not be most efficient or cost effective.

The second approach involves identifying problems in each phase of the project, but not invoking problem resolution until just before the final data set is written. While this approach may be a more efficient use of staff, it may be problematic because of the cascading (domino) effect of failing to correct inaccurate data immediately. The domino effect can produce data errors not readily apparent to a quality control checking program or a person performing manual quality control checking. If problem resolution is invoked once at the end of the conversion effort, plan on devoting extra effort to detecting errors that may result from inaccurate data introduced earlier in the process.

People may be your most valuable resource in problem resolution. Most problems are not a result of data that have not been converted incorrectly from source documents. Rather, errors are usually the result of data that do not meet other criteria identified for the project. To resolve these errors, it is frequently necessary to involve individuals who have been caretakers of the paper records over the years. These people usually have substantial knowledge regarding the data that may never have been recorded on paper. The more information you can obtain from these individuals, the more accurate your database will be. They may also be good sources of information regarding the best end user applications to develop.

If you are not performing the problem resolution process within the data conversion effort itself, but rather forwarding the questions to a third party (a person or organization), you should establish reasonable turnaround times for the results. Many data conversion projects are completed on tight schedules (and budgets), and you cannot afford to delay the conversion process indefinitely. Work with the individuals who will conduct problem resolution to establish a reasonable time period, and define what action will be taken if responses are not received according to the time frame (such as simply converting default values). End users can certainly update the information if better data are obtained in the future.

Managing Quality Risks

The success of a GIS data conversion project can be measured in many ways. Was the project completed on time? Was it completed within budget? What quality level do the data embody? Schedule and budget are easy to measure. To measure data quality, you must first define the measuring stick you will use.

Documentation of Quality Requirements

Data quality requirements must be defined before the first data set is converted. In particular, keep the following factors in mind and use them to guide the process.

- What level of data accuracy can be achieved given the project's cost, scope, and budget?
- What level of data accuracy is required to meet current and future needs of end users?

After addressing these items, it may be necessary to adjust the scope of your project. You may find that the funds allocated for the project are inadequate to support the level of accuracy required by end user applications. If this is the case, consider changing the initial user applications or limiting the size of the geographic area to be converted.

Data quality requirements will be key criteria for measuring the success of your project. If you set the requirements too low, the users will not have confidence in the data. While the data conversion project may be a success, the GIS project will ultimately fail.

Criteria for exceptions to data accuracy requirements should also be defined up front, if possible. If they are left undefined, you may have to commit substantial effort later to resolve problems and questions that arise during the conversion. Ideally, you should incorporate exceptions into the data accuracy standards that will measure the success of the project. In this way, exceptions will not taint your data.

Set Your Sights

Defining data accuracy requirements correctly is a critical success factor for GIS data conversion projects. You must walk a fine line between providing higher quality than necessary and not giving end users what they need. Setting the standards higher than necessary is a common mistake; avoid it if possible. At a minimum, unnecessarily high standards can cause the following problems.

- Requirements cannot be met within the time frame and budget available. Conversion costs can skyrocket, while data quality levels only improve incrementally. As you exceed 95% accuracy, costs can increase logarithmically, while quality goes up on a straight scale.
- User expectations can become so high that they will never be met, and end users will not be happy with the results.
- It may become extremely difficult to maintain morale among the data conversion staff. While people must be challenged, goals must be reasonable as well.

Setting standards lower than necessary can also cause problems, some of which are listed below.

- The data conversion staff can become complacent, introducing errors that are expensive to correct.
- GIS users may become frustrated and return to using old, manual methods to perform their work.
- GIS users may have inflated opinions of data accuracy. The results of their queries and analyses may drive programs in the wrong direction—perhaps the most serious repercussion of poor data quality.

Providing Adequate Resources

A natural tendency in business is to generate as much revenue as possible with as few resources as possible. This tendency can certainly affect data conversion. While it is important to deliver a project within budget, it is also important to define the project in a manner that can be accomplished using available funding.

The data conversion staff is the core of every data conversion project. Equipment and software are also important, but they mean little if a competent staff is not part of the equation. Hire the best people you can afford, and then support them by providing the necessary training to complete the project. Once these individuals are trained and contributing to the project, the next challenge is to retain them. Provide them with achievable goals, offer regular feedback on their performance, reward them when they do well, and provide guidance and/or additional training as necessary. Your people are your project, and your project will become your reputation. Give them your best.

Hardware and software are also important elements in GIS data conversion. Some organizations have a tendency to reallocate resources based on politics or one individual's perceived power. Try to avoid these pitfalls. Inadequate resources can quickly sideline the project.

Resolve the Quality versus Schedule/Budget Conflict

In well-planned GIS data conversion projects, the schedule and available budget are adequate to convert all data at the required accuracy level. However, data conversion projects seldom go exactly as planned. When problems arise, it is tempting to lower the level of quality control in order to correct budget or schedule problems. While the project may be praised because it was completed on time and within budget, resulting data quality problems could haunt you for some time. Instead of lowering quality, consider the following suggestions to bring a project back on schedule and budget.

- **Consider more training.** While they may be getting the job done, additional training may teach them more efficient ways to perform certain tasks.

- **Revisit manual procedures.** Could the programming effort required to automate them be offset by a savings in labor costs? Automating processes can also improve quality control by reducing errors.
- **Have an outsider review the process.** A qualified individual may identify redundant processes that you can eliminate without sacrificing quality.
- **Consider incentives.** Instituting a production quality incentive program may be worthwhile, but be careful. These programs are difficult to design in a manner that is fair to all participants, and the potential for abuse is high. The cost of the program also should be carefully evaluated against projected savings.

The manager of a data conversion project is usually tasked with completing the project on time and within budget. Unfortunately, the emphasis on quality can wane at times. It is up to the data conversion manager to assume this responsibility and balance quality against the schedule and budget. These three factors are equally important in order for the final GIS conversion effort to succeed.

Appendix A: Glossary

Absolute spatial accuracy—Placement of features relative to their true positions on the Earth's surface.

Acceptance criteria—Criteria used to evaluate converted data and determine whether the data meet project standards. See also *Validation.*

Accuracy—A measure of how closely data represent reality.

Aerial photographs—Photographs taken of the Earth's surface with a camera mounted on an aircraft.

Aerotriangulation—Establishing coordinates for points visible on aerial photographs by calculating the distances and angles from other apparent points with known coordinates.

Almanac—A file containing orbit information on all satellites, clock corrections, and atmospheric delay parameters. The file is transmitted by a *GPS* satellite to a receiver on the ground, where it can be downloaded to help determine positions of locations on the ground.

Analog stereoplotter—A device for creating maps from aerial photographs that uses mechanical and optical technologies.

Analytical stereoplotter—A device for creating extremely accurate three-dimensional models from aerial photographs. Device functions are controlled by software.

Analytical triangulation—Populating a ground control network to ensure sufficient *control points* for stereopairs generated for *photogrammetry.*

Annotation—Map text.

Area feature—See *Polygon.*

Attribute data—Data representing the properties of an object. Examples include the voltage carried by a power line, the infiltration coefficients of a particular soil type, or sales within a region.

Automated conversion—Converting maps and drawings into digital form in a manner that minimizes or eliminates human involvement.

Automated mapping—Mapping using various computer technologies.

Base map—Reference data that supply context for an area of interest. Base data are frequently shared among multiple end users (e.g., streets, addresses, and elevation contours). Also known as a landbase.

Base station—A receiver set up at a known location specifically to collect data for differentially correction *rover* files. The base station calculates the error for each satellite and, through differential correction, removes *selective availability* and improves the accuracy of the roving GPS receiver positions collected at unknown locations. Also called a reference station.

Bit map—Gridded pattern of bits used to generate a raster image.

Blitzkrieg data conversion—Data conversion involving an intense and concentrated effort on the part of a project team to initiate and complete a project as rapidly as possible.

Bulk loading—Loading existing data files into a database.

CAD—See *Computer aided design.*

Cartography—The discipline of creating maps.

Cell—In GIS, the basic element of spatial information in the raster data model. In organizational management, a unit with an agency or department independent of the department which has its own budget, schedule, and personnel dedicated to a given project.

Centroid—A point within a polygon used to identify it.

COGO—See *Coordinate geometry.*

Compilation scale—The scale at which a map is converted.

Computer aided design/drafting/drawing (CAD)—Technology related to GIS used primarily for drafting and design in engineering and architecture. Some GIS systems run on a CAD platform. Also called CADD.

Conceptual database design—Defining conceptual data organization, requirements, and associations for a GIS apart from hardware and software considerations.

Consortium—A voluntary agreement among independent entities to work together cooperatively and share resources for the common good of participants. By strict definition, consortium undertakings are beyond the resources of a single member.

Contour—A map line representing imaginary lines of constant elevation.

Control points—Points with known coordinates, which thus become reference points.

Conversion—Translating diverse attribute and feature data (in hardcopy, visual, and digital form) into a cohesive, compatible digital format for use in GIS.

Coordinate geometry (GOGO)—Mathematical entry of bearing and distance data; commonly used to construct extremely accurate parcel base data.

Coordinates—The locations of points according to X, Y, and Z axes.

Coordinate system—A specific system using linear or angular designations to determine the precise location of points on the Earth's surface. See *State plane coordinate system.*

Data—Values measured and recorded through various processes; the foundation of GIS.

Database—A computer system for storing groups of interrelated data sets according to a specific method of organization or structure. See also *Conceptual database design* and *Physical database design.*

Database management system—A specific system for organizing data sets that enables users to query and manipulate tabular data stored in the system.

Data format—The model used to represent, store, and manipulate geographic data on a computer. Common GIS examples include *raster* and *vector* data formats.

Data layer—The GIS approximation or equivalent of a hardcopy map overlay.

Data quality—The umbrella term encompassing data completeness, accuracy, and integrity.

Data set—A collection of data of a single type in hardcopy or digital form.

Data source—Any form of information to be converted for a GIS database, including (but not limited to) hardcopy maps and plots; digital map data; aerial photographs; map symbology and annotation; database records; and tabular information.

Datum—A reference system used to describe the surface of the Earth. Coordinate systems used to survey point locations on the Earth are linked to a datum. For North America, two datums exist, the North American Datum of 1927 (NAD 27) and NAD 83.

DCW—See *Digital Chart of the World.*

DEM—See *Digital elevation model.*

DGPS—See *Differential correction GPS.*

Differential correction GPS (DGPS)—The process of correcting GPS positions at an unknown location with data collected simultaneously at a known location (see *Base station*). Differential correction can occur in real time or in post-processing. In post-processed DGPS, the base station logs the measurements in a computer file so rover users can differentially correct data. In real-time DGPS, the base station calculates and broadcasts the error correction for each satellite as received, permitting rover users to use differentially corrected data immediately. Also see *Rover.*

Digital Chart of the World (DCW)—A worldwide vector data set.

Digital elevation model (DEM)—The term used by the U.S. Geological Survey and other organizations for a regular grid of elevations used to generate shaded relief maps and other GIS applications. Similar terms include *digital terrain model* and digital terrain elevation data.

Digital line graph (DLG)—Vector data produced by the U.S. Geological Survey (USGS) in digital form corresponding to USGS topographic quadrangle maps. DLGs preserve spatial relationships among map features and include attribute information. Also describes several file formats (standard, optional, and graphic) used by the USGS and other federal agencies.

Digital orthophotograph—A photograph derived from aerial photography and corrected to depict features in their true geographic positions. Correction, or rectification, includes removing distortions caused by camera tilt (displacement) and terrain.

Digital raster graphic (DRG)—Raster files containing images of maps; produced by the USGS and other government and commercial sources.

Digital terrain model (DTM)—A portion of the Earth's surface viewed three-dimensionally.

Digitize—Converting hardcopy source maps and map sheets into digital format by tracing.

Digitizer—A tablet or table used to create vector linework from source documents. A source map is taped to the table, and an operator follows the lines and clicks a puck at line vertices. Also refers to the equipment operator.

Displacement—Shifting horizontal placement of objects in a two-dimensional photograph or sensor image.

DLG—see *Digital line graph.*

Dots per inch (dpi)—Method for measuring scanner, printer, and plotter resolution. In general, the higher the dpi, the better the resolution.

Double key data entry—The process of having two operators key in identical attribute data. The two data sets are then compared for accuracy.

dpi—See *Dots per inch.*

DRG—See *Digital raster graphic.*

Drum scanner—Scanner with the light source and camera lenses placed within a rotating drum. Documents and images are mounted on the drum.

DTM—See *Digital terrain model.*

DXF—Exchange format for AutoCAD software, often used to import vector data into and export them out of GIS; not well suited for converting attribute data.

Earth Science Information Center (ESIC)—The U.S. Geological Survey office that provides information about and sells data, maps, and publications produced by the USGS and other federal agencies.

Edge matching—Identifying common points, lines, or symbols on adjacent maps, drawings, or images.

Electronic cooperative—Two or more entities who enter a formal arrangement to share resources across a digital network.

End user—An individual or group who is the primary recipient of a project's output.

Entity—A component of the relational data model. An entity is an idealized representation of a person, place, event, object, or concept about which the user wishes to maintain data.

ESIC—See *Earth Science Information Center.*

Export—The process of moving data out of one computer system for the purpose of *importing* them into another.

External conversion—Data conversion performed by an external conversion vendor, agency, or consultant.

Feature—The basic geographic element in a GIS.

Feature attribute—A nonspatial characteristic of a feature (e.g., type of manhole or water valve).

Federal Geographic Data Committee (FGDC)—A federal government group coordinating spatial data collection.

Feed scanner—The preferred scanner for GIS applications, the feed scanner works by moving the document over fixed camera lenses and a light source. Feed scanners can accommodate large documents such as hardcopy maps. Also known as a direct feed or roller feed scanner.

FGDC—See *Federal Geographic Data Committee.*

Field inventory—Physical features of interest are visited and inventoried at their actual locations. Spatial coordinates and attributes may be collected during a field inventory.

Flatbed scanner—A scanner that works by moving the camera lenses and light source below a flat plate of glass on which a document is placed.

Gantt chart—Used to graphically display the contrast between projected and actual costs and schedules.

GBF/DIME—See *Geographic Base File/Dual Independent Map Encoding.*

Generalization—The process, using manual or software means, to reduce the number of *nodes* in a *line.*

Geocode—Determining the position of geographical objects (landscape or cultural features) relative to a standard reference system or grid normally using addresses.

Geodetic controls—See *Control points.*

Geographic Base File/Dual Independent Map Encoding (GBF/DIME)—The data set preceding the Census Bureau's TIGER/Line files. See also *TIGER.*

Geographic coordinate system—The coordinate system that assigns unique coordinate pairs to each location on the surface of the Earth.

Geographic data—See *Spatial data.*

Geographic information system (GIS)—A system of hardware, software, and equipment to capture, store, manipulate, analyze, and display spatially referenced data.

Geographic Names Information System (GNIS)—Official database of U.S. geographic names; available from the U.S. Geological Survey.

GeoTIFF—An extension to the tagged image file format (*TIFF*) that includes cartographic and georeferencing information.

Global positioning system (GPS)—A system of orbiting satellites used to determine location in three dimensions. The *NAVSTAR* constellation of satellites transmits radio signals identifying satellite locations, which are used to determine a position on the Earth. Managed by the U.S. Department of Defense and available for civilian use, albeit with a degraded signal.

GNIS—See *Geographic Names Information System.*

GPS—See *Global positioning system.*

Grid—A linear network of rows and column whose most basic element, cells, contain data; the format for raster data.

Heads-up digitizing—Digitizing a map by tracing relevant features on-screen with the cursor.

Horizontal control datum—A geodetic reference for horizontal control surveys, composed of five elements: latitude and longitude, azimuth, and two constant values used in connection with the reference spheroid.

Horizontal control point—A survey station with known coordinates.

Hydrography—Mapping watercourses and bodies of water.

Hypsography—The area of topography concerned with elevation or relief.

Import—The process of loading data from an external source into a target system.

Incremental data conversion—Step-by-step approach characterized by periods of inactivity at one or more points in the life of a project.

Information—The result of data analysis. Information is data in a useful form.

Internal conversion—Data conversion conducted by existing staff in house.

Keyboard entry—Operator-controlled entry of attribute data into a computer data file; used in *single key* and *double key data entry* and entering *coordinate geometry.*

Landbase—See *Base map.*

Landsat—Remote sensing satellites managed by the National Aeronautical & Space Administration to collect data regarding natural resources on the Earth.

Layer—Groups of related data within a GIS that can be overlaid in any fashion to yield distinct spatial analyses.

Line—A basic element in *vector GIS,* defined by at least two pairs of XY coordinates.

Linear feature—A feature represented most appropriately with lines.

Location—Describing the position of a physical or cultural feature on the ground in terms of coordinates.

Mandate—A clear directive from a recognized authority for particular action.

Map—Graphic depiction of landscape or cultural features on the surface of Earth using a well established system of symbology and/or photographic imagery. Maps must use projections, indicate the reader's orientation or view, and adhere to established scales.

Map feature—An object or feature depicted on a map.

Map projection—A method for representing the curved surface of the Earth on a flat plane. See also *Universal Transverse Mercator.*

Map scale—The relationship, often expressed as a ratio (1:200,000), between map distance and the same distance on the ground.

Metadata—Data about data that often contain information such as the method of collection or acquisition and the accuracy of a given data set. Metadata allow detailed information about a data set to be reviewed prior to obtaining the data. The Content Standards for Digital Geospatial Metadata is a federal standard for documenting metadata.

Milestone—A specific accomplishment within a project that is easily identified, occurs at a logical interval, and fits into a recognizable context.

Monumentation—The process of creating a permanent marking of public land survey to precisely establish the location of the surveyed area.

NAD 83—See *North American Datum of 1983.*

National Imagery and Mapping Agency (NIMA)—The federal agency that produces maps and data for defense and intelligence purposes; succeeds the Defense Mapping Agency.

National Map Accuracy Standards (NMAS)—Specific horizontal and vertical accuracy guidelines established by the *U.S. Geological Survey* to ensure maps adhere to common definitions of accuracy.

NAVSTAR—An acronym formed from NAVigation Satellite Timing And Ranging; the constellation of 24 satellites used in *GPS.*

Needs assessment—A ranked, formal compilation of needs for a specific group.

NIMA—See *National Imagery and Mapping Agency.*

NMAS—See *National Map Accuracy Standards.*

Node—The term used to in many GIS products to describe a line end point or a point along a line feature.

Normalization—In relational database terminology, the process of rationalizing a data set into simple, stable data structures. In scanning, the process of reducing each feature to a single color.

North American Datum of 1983 (NAD 83)—The horizontal control datum for the United States, Canada, Mexico, and Central America.

OCR—See *Optical character recognition.*

On the fly—Collecting data "as you go," either in the field or during digitizing. Also refers to gathering continuous data.

Optical character recognition (OCR)—Technology that converts scanned text that normally would be recognized as a single image into individual characters.

Output scale—The scale at which a hardcopy map is produced.

Overlap—The area on a map or photograph common to one or more in the same series of images. Overlaps are deliberately created in aerial photography to ensure accuracy of photos in the series.

Overlay—A data layer in GIS used in conjunction with base data contained in a database. To ensure accuracy, overlays must be registered to base data using a coordinate system common to both.

Overshoot—When a line extends past an intended intersection with one or more lines.

Parcel—The basic unit of land, and consequently a fundamental element in many GIS applications.

Photogrammetry—The process of creating maps from photographic images.

Physical database design—Defining the specific framework of a GIS database, incorporating conceptual database design and the requirements of a particular database management system.

Pilot phase—A limited area in a GIS data conversion project used to test the data, techniques, and processes outlined in the project to confirm their appropriateness and make necessary changes.

Pixel—The basic unit of an image (photographic or scanned) or a raster grid.

Planimetric map—A map depicting only horizontal locations of features (e.g., highways, rivers, city streets, and houses).

Point—Usually defined as a location with zero area. Often physical objects (e.g., oil wells) are conceptualized as points.

Polygon—Spatial data representing a region or area.

Positional accuracy—How dependable the placement of cartographic features are on a map relative to each other and the ground.

Precision—A measure of how well data meet a certain standard. In database design, it can refer to the number of significant figures to which data are stored.

Primary data—Data collected directly from the field.

Project champion—An individual who nurtures a project and ensures that support is maintained at all levels of an organization.

Puck—A hand-held device used to digitize features.

Quadrangle—A map or data set bounded by four lines of latitude and longitude. Used for many U.S. Geological Survey maps and data sets.

Quality control—The sum of procedures implemented to ensure data are converted accurately in GIS.

Raster data—Spatial data that represent physical features in a fixed grid of rows and columns. Each cell (where rows and columns intersect) contains a single attribute.

Raster GIS—A GIS comprised of one or more grids of raster data. Raster GIS applications can involve extremely large data files, but they are well suited to spatial analysis and constitute the logical format for remotely sensed data. To the present, raster GIS applications have been common in natural resource management.

Registration—Aligning GIS overlays or hardcopy maps using common coordinate points.

Relational database—A method or structure for storing diverse and independent groups of data or whole databases that enables computer users to transparently manipulate, query, and retrieve diverse data without regard to their specific structure.

Relational data model—A formal, mathematically defined method of representing data in simple, stable data structures.

Relationship—A component of the relational data model that establishes an association between the instances of one or more entities. Relationships allow instances of one entity to be linked with associated instances in other entities.

Relative spatial accuracy—The placement of features relative to the base map data and other features in the GIS.

Remote sensing—Gathering information from a distance.

Resolution—The sharpness of a photographic or scanned image, commonly discussed in terms of *dpi.*

RMS—See *Root mean square.*

RMSE—See *Root mean square error.*

Root mean square (RMS)—A method of measuring the dispersion of a data series around truth. 1RMS includes approximately 68 percent of the data points occurring within the stated distance from truth. 2RMS includes approximately 95 percent of the data points occurring within the stated distance from truth. *Selective availability* at 100m is 2RMS; most *GPS* stated accuracy is 1RMS.

Root mean square error—Square root of the sum of the squared errors in a population testing.

Rover—A mobile GPS receiver collecting data during a field session.

Rubbersheeting—The adjustment of features on a digital map, overlay, or image to compel them to fit with other data.

Scale—The relationship between a ground measurement and the measurement on a (usually) hardcopy map. In GIS, a scale is often assigned to data to indicate their accuracy or the scale of sources used to create the data.

Scan line—A single *pixel* row of an image.

Scanning—The process of electronically converting hardcopy data to digital form for use and storage in various computer environments. In GIS, data are scanned in raster (or grid) format.

Scope creep—The temptation to add another attribute, layer, or other element to a GIS data conversion project. Collectively, these elements can doom a project because they require more time, money, staff, and other resources, causing the project to delay or miss milestones.

Scrub—Manual preparation, verification, and correction of hardcopy map data prior to digitizing.

SDTS—Spatial Data Transfer Standard, FIPS PUB 173, August 1992.

Secondary data—Data derived from existing source documents or other sources.

Selective availability—The artificial degradation of a satellite signal by the U.S. Department of Defense. The error in position can be up to 100m. However, *differential correction GPS* techniques can correct the error.

Side lap—The area where adjoining aerial photographs overlap.

Silver grain, photographic—Silver salts (chloride, bromide, or halide) suspended in a gelatin varnish to form the emulsion in film.

Single key data entry—The process of scrubbing and keying in attribute data once.

Spatial analysis—The hallmark function of computerized GIS applications. The process of using various analytical techniques to assess and compare the geographic dimensions of data. Software packages that lack such functionality are not considered GIS.

Spatial data—Data describing the geographic location of point, line, or area features. Normally expressed in a single XY pair or a series of such pairs. Geographic location can be expressed in various geodetic or plane coordinate systems (e.g., longitude and latitude, or state plane).

Spatial data transfer standard—A standard to facilitate data exchange between different hardware and software systems without losing information.

Spatial object—The fundamental form of geometric data in a GIS. Terminology varies considerably among systems, but lines, points, nodes, areas, polygons, and arcs are all examples of spatial objects.

State plane coordinate system—A system of grids developed by the National Geodetic Survey for each state.

Stereocompilation—Using photogrammetric instruments and geodetic controls to generate a map from one or more aerial photographs.

Stereodigitizing—Collecting data directly from aerial photographs with precision equipment.

Symbol—A graphic element on a map; it can represent point, line, or polygon features.

Table—Often used in relational database management systems to store entity data.

Tabular data—Statistical data or data stored as a table.

Tagged image file format (TIFF)—A raster file format frequently used for imagery. See also *GeoTIFF*.

Target format—Format in which the target system's data are stored.

Target platform—The hardware that supports GIS software and data.

Target system—The intended GIS system that will make use of the converted data. In simple terms, target systems are composed of hardware, software, and usually a network.

Temporal data—Data that vary over time (e.g., flow in a river, and monthly sales in a region).

Thematic map—A map made for the sole purpose of communicating a theme or showing statistical information.

Three dimensional—The ability to represent length (X dimension), width (Y), and depth (Z) graphically.

TIFF—See *Tagged image file format*.

TIGER—See *Topologically Integrated Geographic Encoding and Referencing*

Topography—The location and shape of features on the Earth's surface, including hydrology, cultural features, and relief.

Topologically Integrated Geographic Encoding and Referencing (TIGER)—A nationwide vector database produced by the Census Bureau useful for many GIS applications. Available to the public in TIGER/Line files.

Topological structuring—Encoding the relationships between point, line, and area features in vector GIS systems.

Topological Vector Profile (TVP)—A profile of the SDTS known as Part 4 of FIPS PUB 173-1, June 1994.

Translator—Software that translates data between different formats. Translators are often necessary to import external data into a GIS.

Trilateration—The measurement of distances to known positions to find an originating location. While the process normally uses control points on the Earth's surface, in GPS trilateration, known positions are satellites orbiting the Earth. Distances are found by computing the travel time of the GPS signal to a location on the surface of the Earth.

TVP—See *Topological Vector Profile.*

Undershoot—When a line fails to intersect one or more lines.

Universal Transverse Mercator (UTM)—A well-known map projection that casts a rectangular coordinate system of the Earth onto a cylindrical surface.

U.S. Geological Survey (USGS)—The national mapping organization in the United States; a valuable source of external data.

USGS—See *U.S. Geological Survey.*

Validation—The stage in a GIS data conversion project when converted data are compared against source documents.

Vector data—Spatial data representing specific objects as points, lines, or polygons.

Vector GIS—A GIS based on vector data, where features are represented by coordinate pairs defining points, lines, and polygons.

Vectorize—To convert raster data into vector data.

Vector product format (VPF)—Format used for the *Digital Chart of the World.*

Vertex—The point at which a line changes direction.

Vertical control datum—A geodetic reference for vertical control surveys (i.e., elevation), usually mean sea level.

Wireframe—Digital view of an object where surfaces are displayed as outlines, and elements behind surfaces are displayed as if the surfaces were invisible.

Appendix B: Contributors

Phyllis Altheide holds B.S. and M.S. degrees from the University of Missouri-Rolla. She began her employment with the U.S. Geological Survey (USGS) in 1983 while still a student. She has helped to develop many cartographic systems, including knowledge based geologic map generation, vector data accuracy testing, and elevation and vector data production systems. Currently, she is a senior computer scientist at the Mid-Continent Mapping Center in Rolla. Since 1992, Phyllis has been the USGS technical leader in the implementation of the Spatial Data Transfer Standard (SDTS). She has performed product mappings for USGS digital data products, coordinated software development, participated in profile development, been recognized internationally as an SDTS expert, provided technical assistance and training, coordinated mass data conversions, and authored journal articles and training materials.

Andrew Coates has been developing database applications for more than 10 years. He currently serves as the professional officer at the University of New South Wales (Australia) School of Civil and Environmental Engineering, where he is responsible for the acquisition, verification, storage, retrieval, analysis, and presentation of hydrological data from the school's experimental catchments. Andrew has extensive experience in data acquisition and transmission, and has published papers on data exchange and integration, particularly between GIS and models.

Kelly Dilks is project leader for the U.S. Army Construction Engineering Research Laboratories (USACERL). She earned a B.A. from the University of Illinois in 1990 and an M.S. from the University of Illinois in 1994. She has worked with U.S. military installations and the U.S. Army Corps of Engineers Districts in all aspects of implementing GIS, including data discovery, data conversion, training, hardware, and software selection. Kelly has worked on metadata and data issues with the Federal Geographic Data Committee to assist the U.S. Army Corps of Engineers in creating its node on the National Spatial Data Infrastructure, and regarding training associated with the metadata standard. Kelly leads a team of researchers working on integrating models, spatial data, and analysis techniques for decision making to support managing military training lands.

Atanas Entchev is the GIS director for Owen, Little & Associates, Inc. (OLA), in Beachwood, New Jersey. Atanas earned an M.S. in urban planning from Rutgers University in 1993, and an M.S. in architecture from the University of Sofia (Bulgaria) in 1984. Atanas is responsible for coordinating GIS services for OLA, including needs analysis, data conversion, custom software development, and implementation. An ESRI-certified instructor, he regularly teaches PC ARC/INFO, ArcCAD, and ArcView classes.

During his employment with the Green Acres Program of the New Jersey Department of Environmental Protection, Atanas helped to develop the program's GIS and was responsible for entering municipal, county, and state open-space inventory data into an ARC/INFO based GIS. As part of this project, Atanas designed unique methods for relating attribute and spatial data.

Atanas is the chair of the Parcel Mapping Group of the New Jersey State Mapping Advisory Committee, where he works toward securing state approval of GIS-developed parcel map layers in lieu of traditional tax assessment maps.

Edgar Falkner received a B.S. in Forestry from Michigan Technological University. As a forester, he was involved in wildlands management with the U.S. Forest Service and Alaska Department of Natural Resources for several years, with significant emphasis on image analysis. He spent three decades in private sector photomapping, the last 10 years of which as a founder, partner, and executive with an aerial survey firm. For five years, Ed was a cartographer with the U.S. Army Corps of Engineers. Ed is certified as a photogrammetrist and mapping scientist (remote sensing) by the American Society for Photogrammetry and Remote Sensing (ASPRS). He is the author of *Aerial Mapping: Methods and Applications* (Lewis, 1995), co-author of the U.S. Army Corps of Engineers *Photogrammetry* manual (1993), and originator of several published scientific papers. Several colleges invited him to guest lecture. At present, he acts as a mapping consultant and writer (fiction and nonfiction).

John Kelly, a senior technical consultant with Convergent Group/UGC Consulting of Denver, has provided GIS planning and implementation guidance and services to public sector clients throughout the United States. John earned an M.S. from the School of Public and Environmental Affairs at Indiana University in 1977. He has guided clients in city and county governments in all phases of GIS implementation, including strategic planning, technical specifications, and vendor evaluation assistance for data conversion vendor procurement, pilot project planning and management, data acceptance testing procedures and management, data conversion management, and applications development management. The majority of John's clients use Environmental Systems Research Institute, Inc., software.

Edward Kura is a GIS specialist with Camp Dresser & McKee, Inc., in Orlando, Florida. He has worked in GIS and data conversion for 19 years. In the past, he has worked on GIS data conversion projects in the areas of parcel mapping, natural resources, land use management, industrial site management, and general mapping applications. With Camp Dresser & McKee, he has managed the GIS data conversion of the water and sewer systems for the City of Chicago. His current assignments include data conversion projects in the areas of storm water management, parcel mapping, and water supply systems.

Bob Lazar is currently a technical development specialist for Rand McNally. Previously, Bob has worked for Applied Geographics, Inc., the Milwaukee (Wisconsin) GIS office, and the U.S. Geological Survey. He received a B.S. in cartography from the University of Wisconsin-Madison in 1979 and attended graduate school at Ohio State University. Bob has 18 years of practical GIS experience in development and implementation of GIS applications, project management, and standards development.

Richard Lewis is president of RLA Communications, a geopositioning dealer for Trimble Navigation, Environmental Systems Research Institute, Inc., Laser Technology, and several differential correction services. He is a certified instructor in global positioning systems and GIS. Contact RLA via its Web page *(www.rlageosystems.com)* or e-mail *(rlacom@ix.netcom.com).*

Adrien Litton has nine years of experience in database design and computer systems administration, and extensive experience with shipboard navigation and communications. In 1990, he joined Environmental Systems Research Institute, Inc., as a data processor and is now the scanning specialist and leader of the Scanning Services Group in the Professional Services Division. Adrien has authored and presented two papers for ESRI's Annual Users Conference Proceedings, and he has led workshops at ESRI user conferences in 1994, 1996, and 1997. He also authored an article for *GIS World* on scanner accuracy. Adrien served with the U.S. Coast Guard from 1981 to 1990, where he achieved the rank of quartermaster second class. He received a Coast Guard Achievement Medal in 1988 and a Coast Guard Special Operations Ribbon in 1986. He performed navigational duties during a scientific expedition to Antarctica in 1984, earning an Antarctic Service Medal.

Marty McLeod is a senior water engineering technician with the City of Riverside (California) Public Utilities Department. He serves as the Water Division's GIS project coordinator, where he has been responsible for data conversion, software development, and implementation since 1991. Marty loves baseball and once tried out for the Los Angeles Dodgers.

Dennis Morgan received a B.S. in engineering technology from Northern Arizona University-Flagstaff in 1973 and is certified as a photogrammetrist by the American Society for Photogrammetry and Remote Sensing. He has spent 25 years working for the U.S. Army Corps of Engineers in St. Louis, Missouri, the last nine years of which as a civil engineer in the Geodesy, Cartography and Photogrammetry Section. Dennis currently oversees technical management of all mapping projects for the St. Louis District and lectures on photogrammetry subjects to provide technical assistance to other districts, as well as federal and state agencies.

Samuel Ngan is an application developer specializing in CAD/GIS development. He earned an M.A.S. in digital mapping and cartography from the University of Glasgow (Scotland) and has taken the graduate course in land information management at Ohio State University. He currently works for Woolpert, a Dayton, Ohio based engineering consulting firm developing mapping/GIS applications ranging from pentop based data collection to public Internet access. He also authored an article for *MicroStation Manager.*

Kevin Struck coordinates GIS data conversion, system administration, application development, and grant writing for Sheboygan County, Wisconsin. He received a B.S. in Geography from the University of Wisconsin-Oshkosh in 1987 and has been involved with project management since 1991. Kevin has worked in the public and private sectors using MicroStation, ARC/INFO, ArcView, and Avenue to facilitate data conversion.

Index

Numerics

A

B

C

D

G

H

I

N

Q

R

S

V

W

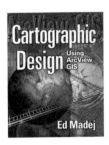

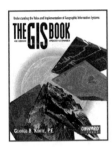

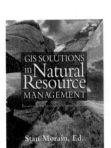

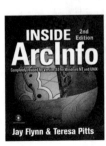

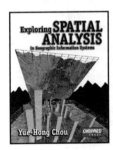

GIS Online: Information Retrieval, Mapping and the Internet

Brandon Plewe

A comprehensive guide to building a Web site based on GIS and mapping technology, including discussion of GIS retrieval and data sharing concepts, motivations for sharing geographic data, plus step-by-step instructions for developing a web site based on GIS and mapping technology.

Order Number: 1-56690-137-5

336 pages, softcover, 7" x 9"

Processing Digital Images in GIS: A Tutorial Featuring ArcView and ARC/INFO

David L. Verbyla and Kang-tsung (Karl) Chang

This book is a tutorial on becoming proficient with the use of image data in projects using geographical information systems (GIS). The book's practical, hands-on approach facilitates rapid learning of how to process remotely sensed images, digital orthophotos, digital elevation models, and scanned maps, and how to integrate them with points, lines, and polygon themes. Includes companion CD-ROM.

Order number 1-56690-135-9

312 pages, 7" x 9" softcover

Focus on GIS Component Software: Featuring ESRI's MapObjects

Robert Hartman

This book explains what GIS component technology means for managers and developers. The first half is oriented toward decision makers and technical managers. The second half is oriented toward programmers, illustrated through hands-on tutorials using Visual Basic and ESRI's MapObject product. Includes companion CD-ROM.

Order number 1-56690-136-7

368 pages, 7" x 9" softcover

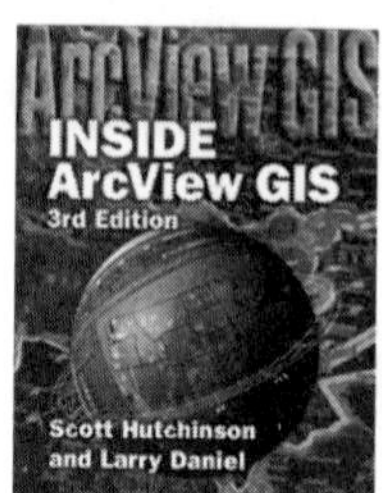

INSIDE ArcView GIS, Third Edition

Scott Hutchinson and Larry Daniel

Written for the professional seeking quick proficiency with ArcView, this new edition provides tips on making the transition from earlier versions to the current version, 3.2, and includes an overview of new extensions. The book also presents the software's principal functionality through the development of an application from start to finish, along with several exercises. A companion CD-ROM includes files necessary to follow along with the exercises.

Order number 1-56690-169-3

512 pages., 7-3/8 x 9-1/8" softcover

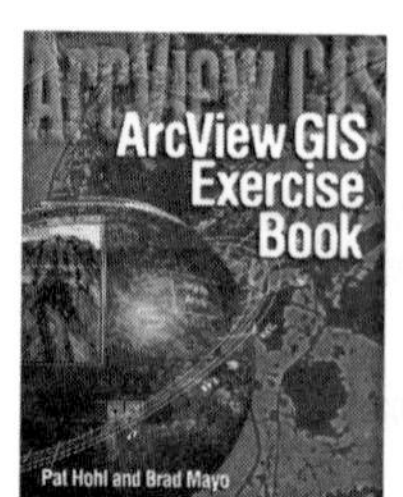

ArcView GIS Exercise Book, Second Edition

Pat Hohl and Brad Mayo

Written to Version 3.x, this book includes exercises on manipulation of views, themes, tables, charts, symbology, layouts and hot links, and real world applications such as generating summary demographic reports and charts for market areas, environmental risk analysis, tracking real estate listings, and customization for task automation.

Order number 1-56690-124-3

480 pages, 7" x 9" softcover

ArcView GIS/Avenue Programmer's Reference, Third Edition

Amir Razavi and Valerie Warwick

This all-new edition of the popular *ArcView GIS/Avenue Programmer's Reference* has been fully updated based on ArcView GIS 3.1. Included is information on more than 200 Avenue classes, plus 101 ready-to-use Avenue scripts—all organized for optimum accessibility. The class hierarchy reference provides a summary of classes, class requests, instance requests, and enumerations. The Avenue scripts enable readers to accomplish a variety of common customization tasks, including manipulation of views, tables, FThemes, IThemes, VTabs, and FTabs; script management; graphical user interface management; and project production documentation.

Order number 1-56690-170-7

544 pages, 7 3/8" x 9 1/8"

ArcView GIS Avenue Scripts: The Disk, Third Edition

Valerie Warwick

All of the scripts from the *ArcView GIS/Avenue Programmer's Reference, Third edition*, with installation notes, ready-to-use on disk. Written to Release 3.1.

Order number 1-56690-171-5

3.5" disk

ArcView GIS/Avenue Developer's Guide, Third Edition

Amir Razavi

This books continues to offer readers one of the most complete introductions to Avenue, the programming language of ArcView GIS. By working through the book, intermediate and advanced ArcView GIS users will learn to customize the ArcView GIS interface; create, edit, and test scripts; produce hardcopy maps; and integrate ArcView GIS with other applications.

Order number 1-56690-167-7

432 pages, 7" x 9" softcover

INSIDE MapInfo® Professional, Second Edition

Angela Whitener, Paula Loree, and Larry Daniel

Based on Release 5.x

Now updated to the software's latest features and functions, this book continues to set the standard for desktop mapping how-to and reference manuals. Essential software functions are revealed through development of a single application from start to finish. Step-by-step examples, case studies, notes, and tips are also located throughout the book to assist readers in their quest to make optimal use of one of today's most popular desktop mapping applications.

Order number 1-56690-186-3

500 pages, 7" x 9" softcover with CD-ROM

MapBasic Developer's Guide

Angela Whitener and Breck Ryker

MapBasic is the programming language for the popular desktop mapping software, MapInfo Professional. Written to 4.x, *MapBasic Developer's Guide* is a handbook for customizing MapInfo Professional. The book begins with a tutorial on MapBasic elements, the MapBasic development environment, and program building basics. Subsequent chapters focus on customizing and editing of all program components. Companion disk included.

Order number 1-56690-113-8

608 pages, 7" x 9" softcover

GIS: A Visual Approach

Bruce Davis

This is a comprehensive introduction to the application of GIS concepts. The book's unique layout provides clear, highly intuitive graphics and corresponding concept descriptions on the same or facing pages. It is an ideal general introduction to GIS concepts.

Order number 1-56690-098-0

400 pages, 7" x 9"

GIS: A Visual Approach Graphic Files

This set of 12 disks includes 137 graphic files in Adobe Acrobat, plus the Acrobat Reader. Corresponding with chapters in *GIS: A Visual Approach*, nearly 90% of the book's images are included. Available in Windows or Mac platforms. Ideal for instructors and organizations with a large GIS user base.

Order number 1-56690-120-0

Set of disks

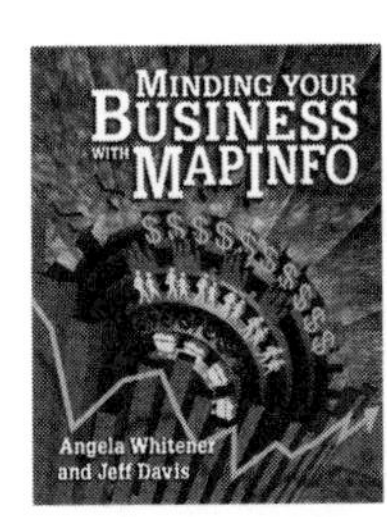

Minding Your Business with MapInfo

Angela Whitener and Jeff Davis

For proficient and prospective MapInfo users, this book offers a conceptual discussion of using desktop mapping and MapInfo products in diverse business activities, including marketing, information systems, and banking, among others. Busy professionals new to desktop mapping can use this book to gain insight into the advantages of mapping business data.

Order number 1-56690-151-0

312 pages, 7 x 9" softcover

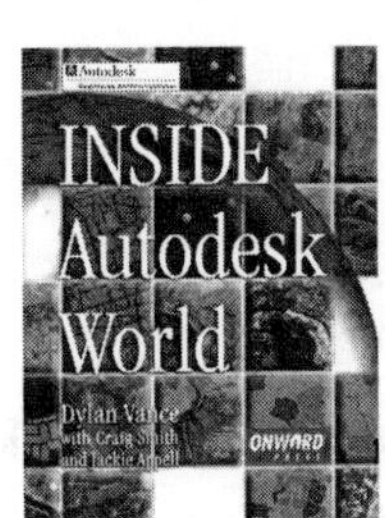

INSIDE Autodesk World

Dylan Vance, Craig Smith, and Jackie Appell

Based on Release 1

Autodesk World is a new tool for managing geographic-based data in a powerful, flexible Microsoft Windows/Office environment. This book leads the reader through the workings of this new software, including getting started with projects and data sets, integrating data from different sources, manipulating data in attribute databases, displaying data, creating data, querying data, presenting maps on the Web, and printing maps. Advanced topics include geobase creation, customizing World, and creating your own application. CD-ROM includes a 30-day time-out version of the software.

Order number 1-56690-179-0

472 pages, 7 x 9" softcover with CD-ROM

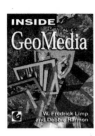